International Perspectives on Social Policy, Administration, and Practice

Series Editors
Sheying Chen, Pace University
New York, NY, USA

Jason L. Powell, Department of Social and Political Science
University of Chester
Chester, UK

The Springer series International Perspectives on Social Policy, Administration and Practice puts the spotlight on international and comparative studies of social policy, administration, and practice with an up-to-date assessment of their character and development. In particular, the series seeks to examine the underlying assumptions of the practice of helping professions, nonprofit organization and management, and public policy and how processes of both nation-state and globalization are affecting them. The series also includes specific country case studies, with valuable comparative analysis across Asian, African, Latin American, and Western welfare states. The series International Perspectives on Social Policy, Administration and Practice commissions approximately six books per year, focusing on international perspectives on social policy, administration, and practice, especially an East-West connection. It assembles an impressive set of researchers from diverse countries illuminating a rich, deep, and broad understanding of the implications of comparative accounts on international social policy, administration, and practice.

Jason Powell

New Perspectives on Health and Social Care

Springer

Jason Powell
Office of the Provost
Crescent College Birmingham
Birmingham, United Kingdom

ISSN 2625-6975 ISSN 2625-6983 (electronic)
International Perspectives on Social Policy, Administration, and Practice
ISBN 978-3-031-25431-4 ISBN 978-3-031-25432-1 (eBook)
https://doi.org/10.1007/978-3-031-25432-1

This Springer imprint is published by the registered company Springer Nature Switzerland AG
The registered company address is: Gewerbestrasse 11, 6330 Cham, Switzerland

Preface

This original book is one of the first to explore contemporary health and social care developments in the occidental world through a conceptual lens that is utilised by Michel Foucault and innovative conceptual tools and differential epistemological theories. In the current socio-economic climate in the UK, for example, health and social care are at a crossroads with implications for how service users' needs are being met or not, how assessment encroaches on power and surveillance and the extent to which service users 'voice' is being translated into the health and social care policy and practices that help them facilitate their independence and, crucially, quality of life. This relentlessly critical book is one of the first to develop new perspectives deriving from Foucault in the application of health and social care policy, theory and practice. The book draws heavily on Foucauldian insights from social gerontology as it is older people predominantly, whilst not exclusively, who access health and social care services; adding a vital and new theoretical perspective to understanding health and social care.

Contents

Chapter 1
Introduction: Perspectives on Health and Social Care

Introduction

There has long been a propensity in matters of health and social care to engage in the ontological and epistemic reductionism of health to its biological and psychological dimensions. Indeed, in Western culture, health and care practices came to be understood in terms of biological science to be only material, and the scientific approach to medicine became overwhelmingly objective, reductionistic and rational. These scientific dimensions primarily are a set of normative 'stages' of body and mind process that position the experiences and representations of service users in Western culture (Gilleard & Higgs, 2000). As the preceding chapter indicated, for the biomedical model, growing old would primarily be a process of inevitable physical and mental 'decline' and of preparation for the ultimate ending: death itself. The paradox, of course, is that the biomedical homogenising of the experience of health which the reliance on the biological and psychological dimensions of health entails is in fact one of the key elements of the (public) dominant 'common-sense' discourses on health and care. A deeper under-standing of health requires, however, that we move beyond common-sense approaches and broaden our view to understand how processes, from the biomedical level of the individual cell to overall society, influence us and in turn are influenced by us as individuals progressing through the life-course.

We also highlighted how health and care was a social category in terms of social constructionism with reference to its definition as both a categorisation and populational construct. The biomedical and social conceptualisations of health and social care demand theoretical interrogation of what Phillipson (1998) calls the 'social construction of health'. The social construction of health is an important process in debunking discourses of 'truth' and can be used as an alternative to narrow medical narratives (Powell, 2002); it includes how 'norms' or pervasive attitudes materialise – from basic biomedical functions to sophisticated and complex social/cultural

J. Powell, *New Perspectives on Health and Social Care*, International Perspectives on Social Policy, Administration, and Practice,
https://doi.org/10.1007/978-3-031-25432-1_1

structures including educational, political and religious institutions, the arts, customs, morality, ethics and law.

Social constructs are enormously powerful in determining individual and collective identity because they answer profound existential questions: Who am I? Where do I belong? What do I do? How do I do it? Where am I? Why am I? Who are you? Why are you? Those in power present ontologically arbitrary social constructs as 'the way life actually is', that is, as reality. For example, biomedical sciences have been important in this process in shaping reality regarding the health process. The social constructions of health assert that health has no existence independent of social interaction and power relationships in society; they are not grounded in 'nature' as is biomedical gerontology under the auspices of biological-psychological gerontological knowledge (the meaning of which is itself socially determined). Indeed, the constructedness of health is made invisible by the normal workings of social life, so that it appears natural rather than artificial. Social constructionism is therefore about helping individuals to stand outside their own prejudices and see the notion of the 'other' in order to explain and understand what may be happening in the social world, given that reality is increasingly seen as fragile, shifting and coexisting with different 'realities'. Further, embracing the social construction of health can help individuals blame themselves less for their 'problems' and strive to change limiting biomedical discourses of human behaviour.

The promise of a sociology of health then could transcend narrow biomedical explanations; it might take a cue from C. Wright Mills's tour de force *The Sociological Imagination* (1959). Mills suggested that the promise and responsibility of the discipline of sociology lie in giving individuals the conceptual tools to make distinctions between 'personal troubles' and 'public issues'. Social theorists can make this distinction if they have a social context and a sense of history from which to understand personal experiences.

The ability to *shift* perspectives, to analyse an experience or an issue from many levels of analysis and to see the intersection of these levels and mutual influence is the heart of the sociological imagination. If we develop a new understanding of our own attitudes about health because we learn about how societies construct meanings of age, then we will have experienced the 'sociological imagination'. As Mills (1959, 5) points out: 'No social study that does not come back to the problem of biography, of history and of their intersections within a society has completed its intellectual journey'.

With these 'tools for thinking', Mills focuses our attention on the broad social structures that shape our personal stories. Equipped with this understanding, we can go on to understand how 'by the fact of our living, we contribute, however minutely, to the shaping of our society and the course of history, even as we are made by society and its historical push' (Mills, 1959, 4). In order to understand health and social care, we need to be aware of how our personal knowledge is shaped by ourselves and by society as a whole.

Drawing from this definition and process of the 'sociological imagination' and the interplay between social context and individuals, the sociology of health and care can be defined as the systematic study of the taken-for-granted assumptions that are socially constructed by society and filter through to shape personal attitudes about health and care in everyday life. The sociology of health is also the study of the relationships between institutions and individuals in society. The sociology of health and care provides an analytical framework for understanding the interplay between human lives and changing social structures. Sociologists are interested, therefore, in how society works, the arrangements of its structure and institutions and the mechanisms of its processes of change.

Hence, the sociology of health is important in examining the interdependence between health over the life-course as a social process and societies and groups as classified by age. The field of social gerontology is firmly related to health and social care and contributes to it through re-formulation of traditional emphases on process and change and on the multi-dimensionality of sociological concerns as they touch on related aspects of other social science disciplines.

Further, the field of social gerontology is concerned with both basic sociological research on age and its implications for social theory as well as policy and professional practice. Nevertheless – and it is almost an embarrassing statement to make – as a field of study, mainstream sociology has not been interested in the sociology of health. As Powell (2002) point outs, in male-dominated sociology, the study of health has been seen as one of the lowest-status areas of all. Despite the poverty of theory in social gerontology in recent years (Bengston et al., 1997), this chapter interrogates and problematises the ways social theories have arisen in relation to interpretations of health, with particular focus on American and British social gerontology.

This chapter focuses on highlighting the major theoretical schools in social gerontology in relation to health and social care by addressing functionalist gerontology, political economy of old age and feminist gerontology. It is arguable that these three theoretical models of health are dominant in modernist constructions of knowledge of health, but they remain important reference points about the nature of health and its structural, interpersonal and esoteric constructions.

This book explores both conceptual and theoretical issues that impinge on understanding health and social care in (post)modern society. It analyses how knowledge formation of health and social care, with particular reference to 'health and social care' in contemporary Western society, is socially constituted and positioned by powerful taken-for-granted assumptions. These assumptions have provided a power/knowledge base for biomedical disciplines, the legitimacy of political-economic discourses and the practices of professional experts.

Given this objective, the book explores theoretical issues about the ontological status of knowledge, examining and moving beyond conventional biomedical philosophical dualisms of 'health and social care as decline'/'decline as health and social care'. Indeed, Michel Foucault's (1977a, b, c) notion that we are governed by means of social processes by which we affirm our freedom is highly pertinent. The informed use of 'thinking tools' from social theory is part of the epistemic aim to debunk and shatter those taken-for-granted assumptions about constructions of health and social care, its biomedical contestations and theoretical relationships of macro/micro, local/global and object/subject to issues of culture, surveillance, governmentality, power relations, trust, narrative, power/knowledge and risk.

Indeed, the book addresses a fundamental issue: can 'meta-theories' provide an effective analysis of health and social care that is radically different from modernist 'grand narratives' as epitomised not only by biomedical models of health and social care but also by mainstream social theories of health and social care inclusive of modernist epistemes of functionalist theories of activity and disengagement, feminist health and social care and the political economy of care health and social care.

The critically acclaimed sociologist C. Wright Mills, in *The Sociological Imagination* (1959), powerfully illustrates that the sociologist seeks, first, to understand the relationship of personal troubles and public issues and the intersection of biography and history; then, these polarities, and the ways in which sociologists address them, define the central problems of social theory in modernity. Mills further argues that the role of the sociologist is to reveal the connections between what is going on in the world and what may be happening within ourselves – in other words, 'to grasp history and biography and the relations between the two within society'. Mills believed that sociologists develop a quality of mind that enables them to show individuals how their own private troubles may be linked to features of the 'public world' of modernity.

Modernity itself was both a 'war of liberation' and a 'war against mystery and magic' (Bauman, 1992, x). It was thought that the world should be despiritualised and instead replaced with 'Big Science' (Bauman, 1992, x). The 'master narrative' of 'progress' became the new 'God', 'rational' replaced 'irrational' and truth became the new 'mission'. In modernity the ultimate goals of progress and the mastery of nature became assumed and unquestioned, just as God had gone unquestioned in religious faith (Bauman, 1992, xiv). Zygmunt Bauman has suggested that a distinguishing characteristic of modernity was the desperate search for a structured world. In which a search for social facts, such as the ways in which individuals come together in societies through a 'social contract', or the ways in which 'social facts' may determine individual behaviour, continue to be debated along with newer concerns with levels of analysis, freedom, determinism and morality. Indeed, in recent years, the German philosopher Jürgen Habermas (1992) has presented us with the doctrine of dialogical morality in modernity, which suggests that the identification of freedom must be arrived at collectively by all those who work in health and social likely to be affected by its adoption.

As a rival to the modernist preoccupation with macro grand narratives of 'progress' and 'reason', postmodernism became increasingly fashionable in its epistemological war against grand narratives. It brought with it an 'anything goes' social context 'which renounces purity, mastery of form and elitism and is more playful, ironic and eclectic in style'.

Postmodernism not only affects culture but also impinges on a range of developments in all areas of Western society, such as politics, industry and the media and the rise of social movements, ultimately leading to the development of a postmodern world characterised by 'fragmentation, multiplicity, plurality and indeterminacy'. If this characterisation is correct, it has had strong implications for the Enlightenment aims and values of modernist approaches of social theory. It is the acceptance that modern aims of universalism are futile and the recognition of 'pluralism of cultures, communal traditions, ideologies, "forms of life" or "health and social care games"' (Bauman, 1992, 102). Whereas for modernists such as Habermas (1992), universalism of collective action is central to determine social change in a social world, for postmodernism (Powell & Longino, 2001), microprocesses of individualisation, subjectification and risk are part of fragmented social spaces, through which identities are formed and performed, that will run counter to macro processes.

Henceforth, the interpretation of the positioning of health and social care as macro- and fixed explanations or micro- and fluid existentialism lies at the heart of modern-postmodern debates within contemporary culture that are plugged into a contestation of 'health and social care': how is health and social care defined? What constitutes the health and social care process? How relevant is science in its understanding? This demands interrogation from transparent theoretical perspectives.

Clearly, what is striking is that one of the major problems of understanding aging in recent years is that it has evidently not been directed by social theory (Powell & Longino, 2002). Indeed, theoretical innovations in health and social care have, according to Vern L. Bengston and colleagues (1997), lagged well behind other sociological master narratives of race, class and gender. Linda George (1995, 77) consolidates this by claiming that gerontological research has been 'theoretically sterile'. For example, in a study of published refereed chapters in eight leading North American health and social care journals (as listed in the Social Sciences Citation Index) from 1990 to 1995, Bengston and colleagues (1997) claimed that 80 per cent of the refereed chapters lacked a theoretical framework. Bengston and colleagues (1997) further argue that social health and social care, a dimension of gerontology, has been 'itty-bitty' in its development of theoretical frameworks to assist an understanding of an important issue such as health and social care: 'it is intellectually irresponsible for a program of research to proceed without a theory' (Bengston et al., 1997, 73).

Ironically, part of the problem about lack of social theory in the study of health and social care relates to the field of health and social care itself. Health and social care as a scientific discipline has been dominated with a preoccupation with biomedical sciences and its constituent elements of 'decline' models of biology and psychology (Estes & Binney, 1989; Longino & Powell, 2004). Health and social care based on social theories of health and social care sees health and social care as a socially constructed category with differential epistemological prisms: for example, functionalism and feminist health and social care. However, while both definitions are fundamental to the complexities of health and social care in the social world, the theoretical interpretations of health and social care are in their 'infancy' (Estes et al., 2003).

If we take the disciplinary dimensions of health and social care, we can illuminate both the relevance and importance they have for understanding the social constructions of health and social care, as well as raising questions about the development of theorising from wider social theory. We can suggest that health and social care has two focal points in its broad conceptualisation: (1) health and social care as science and (2) health and social care as social theory.

Health and Social Care as 'Science'

One of the cornerstones of modern health and social care has been a belief in 'science' and 'progress' as constituting 'discursive practices that give rise to epistemological figures, sciences and possibly formalised systems' (Foucault, 1972, 191). Such powerful discourses systematise networks of ideas about the 'nature' of health and social care, the reasons for particular behaviours and the ways individuals may be classified, selected and controlled. The 'project of modernity' has inspired the disciplinary development of health and social care to reconstruct health and social care on the basis of individual abilities, needs and functions. The health and social care subject is constructed as an object of knowledge and as a seeker of that knowledge. This tension gives health and social care its character. On the one hand, it produces the subject as an empirically verifiable entity, and on the other hand, it produces a critical inquiry into the empirical conditions that justify the existence of subjects. Hence, we get a developmental duality of an empirical and transcendental entity. Indeed, knowledge formation in Western health and social care in particular is modelled and characterised by quantitative, positivist and scientific discourses (Katz, 1996; Powell, 2001b; Biggs & Powell, 2001; Powell & Biggs, 2000; Longino & Powell, 2004). Traditional forms of medical-scientific expertise and knowledge under the rubric of the 'biomedical model' have attempted to foster understanding of health and social care and construct 'worldview' 'truths' about health and social care that are perceived as 'master narratives' (Powell, 2001b). Master narratives are dense forms of discourse containing alleged universal truths, totalising views and symbols with which to explain and understand almost every aspect of social life. These narratives play a central role in the construction of physical and symbolic boundaries, and it is through them that social groups come to know and understand the social world and constitute the classification of perceived health and social care identities (Biggs & Powell, 2001).

Such scientific narratives construct a particular positioning of health and social care. For example, Bytheway (1995) suggests the notion of 'growth' is a central scientific discourse relating to the true changes to the biological body associated with human health and social care. Growth is seen as a positive development by biologists (Bytheway, 1995) in that a baby grows into a child who grows into an adult, but then, instead of growing into care health and social care, the person declines. This scientific sanctioned perception is that growth slows when a person reaches 'care health and social care' and is subsequently interpreted as decline rather than as change, which is taken for granted with earlier life-course transitions. These narratives can be understood as textual formations composed of a complex set of truth codes and conventions through which scientifically grounded and privileged forms of true knowledge strive to present their worldviews as universal and hence valid for the whole of society.

Hence, health and social care is a site upon which power games are played out mainly through such narratives. As Biggs and Powell point out: '. . . narratives are not simply personal fictions that we choose to live by, but are discourses that are subject to social and historical influence' (2001, 113).

Coupled with this, the totalising views of biomedical science, as Longino and Powell (2004) have pointed out, make a particular representation of the world seem so natural that health and social care, for example, cannot be imagined as an alternative discourse except as an abnormality that can be understood only through biomedical science.

However, the objectivity and neutrality of scientific knowledge expressed in broad approaches such as the biomedical model can be understood as rhetorical surfaces that obscure subterranean, politically structured orders. Coupled with this, it has become increasingly clear that a 'scientific' knowledge formation lacks appropriate health and social care language for practitioners for discussing theoretical and philosophical concerns or appreciating their historical and contemporary cultural contexts (Kastenbaum, 1993). Therefore, as Katz (1996) has identified, health and social care is not just a scientific process, and for this reason, it cannot be singularly analysed via 'disciplines' such as 'biomedical health and social care'. Furthermore, according to Katz (1996, 55), the effects of the 'decline' analogy can be seen in the dominance of biomedical arguments about the physiological 'problems' of the 'health and social care body'. Indeed, the master narrative of biological decline hides the location of a complex web of intersections of negative ideas constituting a culture of health and social care (Powell, 2001). Foucault suggests that the surveillance of bodies was central to the development of modern regimes of power and knowledge or what Estes (1979) calls 'health and social care enterprises' – which are institutions created health and social care the 'problems' of service users, from social services to nursing care through to social work with vulnerable people (Powell & Biggs, 2003; Bengtson et al., 2009). Katz (1996, 27) asks: 'How . . . was the figure of the health and social cared body central to the modern constitution of care health and social care and development of biomedical knowledge?' (emphasis in original).

The biomedical model is a powerful discipline and practice, but for Powell and Biggs, it obscures wider understanding of health and social care: 'It appears . . . that established and emerging "master narratives" of biological decline yet linked through the importance of techniques for maintenance . . . via medicalized bodily control. However, this focus on medicalization . . . has tended to obscure another . . . discourse on health and social care . . . the association between care health and social care and social theory [emphasis in original]' (2000, 95).

Health and Social Care as Social Theory

The theoretical concerns of positivist biomedical disciplinarity have significant implications for the social discourses that impinge upon the social construction of health and social care. How have these social discourses and policy implications been interpreted via social theories of health and social care?

A striking feature of theoretical frameworks of health and social care is that the majority of studies are relatively small in scale (Moody, 1998). Despite functionalist

(Cumming & Henry, 1961), Marxist (Phillipson, 1982, 1998; Walker, 1981), feminist (Arber & Ginn, 1995) and 'postmodern' (Featherstone & Wernick, 1995; Gilleard & Higgs, 2000) forms of analysis, health and social care has remained 'theoretically sterile' in comparison with other social science disciplines such as political science and criminology (George, 1995); as an apparently 'applied' field of gerontological study, it has remained 'itty-bitty' (Bengston et al., 1997) in theorising health and social care. Bengston and Schaie (1999) also claim that theoretical developments in health and social care have at best a limited history (1999, 41). Katz (1996, 42) consolidates this by claiming that the use of health and social care in ways that are informed by the cultivation of wider social theories of postmodernism and Foucauldian studies, for example, is relatively unknown territory.

In order to add and develop the relationship of health and social care and contemporary social theory in an area left 'uncharted', this book contributes to, and strengthens, the critical interconnection between health and social care and contemporary social theories of health and social care body, power/knowledge, discourse and subjectivity and risk. While, for example, postmodernism, Foucauldian theory and risk are challenging other disciplinary fields in social science, mainstream social health and social care is impermeable (Estes et al., 2003). Surrounded by its biomedical paradigms, health and social care fails to recognise that its most imaginative developments come from the critical and cutting-edge theories of those scholars who transcend so-called fixed disciplinary boundaries. It is therefore timely that this book is written when serious questions (Phillipson, 1998; George, 1995; Bengston et al., 1997; Biggs, 1999) are being raised about the limited development of wider social theory in social health and social care.

Despite this, the 'biomedical' study of health and social care has dominated the disciplinary development of health and social care which has masked the historical development in theorising health and social care (Katz, 1996). The biomedical model problematises health and social care as a pathological 'problem' tied to discourses of 'decline', 'dependency', 'decay', 'abnormality' and 'deterioration' (Powell, 2002). Thus, the problem orientation to health and social care is historically configured in biomedical sciences and discourses that specialise in the medical reductionism of health and social care (Powell, 2001a; Powell & Biggs, 2000; Biggs & Powell, 2001). As Foucault pointed out: 'It [is] a matter of analyzing . . . the problematizations through which being offers itself to be, necessarily, thought – and the practices on the basis of which these problematizations are formed' (Foucault, 1977a, b, c, 11).

The biomedical problematisation of health and social care has secreted wider questions of power, inequality and culture, and the growth in 'social aspects' of health and social care has developed as a direct challenge to the authority of biomedical power and knowledge. The purpose of this book is not only to challenge such knowledge formation but also to map out the terrain of evocative modern and postmodern theories and their contestations and insights for understanding health and social care. It has only been in the past few years that social theory has been taken seriously within health and social care (Estes et al., 2003; Bengtson et al., 2009).

Following comments by Bengston and colleagues (2009) on the lack of social theory in health and social care studies, the embryonic state of a sociological analysis of health and social care can be judged from the lack of refinement of the term 'health and social care'. In Western societies, an individual's health and social care is counted on a chronological foundation, beginning from birth to the current point of health and social care, or when an individual has died. Counting health and social care is a social construction because it is a practice underpinned by conceptions of time in regional, national and global spaces (Powell, 2001b), which came to be of increasing importance with the development of industrial capitalism (Phillipson, 1982).

Furthermore, health and social care has three main focal points of interest to its theorisation. First, health and social care and health and social care have a biological and physiological dimension, so that over time and space, the appearances of physical bodies change (Longino & Powell, 2004). Second, the health and social care of an individual takes place within a particular period of time and space. Third, as individuals, society has a number of culturally and socially defined expectations how people of certain health and social cares are supposed to be-have and how they are positioned and classified. 'Care health and social care', for instance, is difficult to define, especially for the state and its institutional branches. For example, for UK's 'Department for Work and Pensions', the legal concept of 'pensionable health and social care' has defined 'care health and social care' at 65 (Biggs, 1999). The Department of Health's National Service Framework defines 'care health and social care' at 50 (Powell, 2001), yet the same UK department states that those people requiring intensive health services such as hospitals have been predominantly those career people health and social cared 75 and over (Health and social care Concern England, 1997; Phillipson, 1998). The British state is uncertain what care health and social care can be defined as, but it is clear that biomedical models of health and social care and their viewpoints do influence societal perceptions of health and social care that impinge on social processes such as life zones of 'health', 'work' and 'retirement' (Phillipson, 1998).

It is particularly apt, then, to attempt to ground developments in a social theory that can be applied to questioning what we understand by 'health and social care'. It is also clear that theoretical perspectives need to be documented and analysed in the light of the triumvirate of social, political and economic transformations in Western society over the past 50 years. Indeed, the book also reflects on the modern/postmodern duality and how this continual provocative debate in social theory impinges on health and social care studies, by paying particular attention to examples drawn from biomedicine, social theory, popular culture, power relations and risk.

Mapping Out the Terrain: Modern and Postmodern Theories of Health and Social Care

What is a theory? A theory asks why a particular analogy is used to explain what is meant by health and social care and how the assumptions contained within biomedicine and policy spaces influence our understanding of the position of career people in contemporary society. Theories of health and social care are important in establishing frameworks for understanding, interpreting and problematising health and social care, how the processes of health and social care are contested and negotiated and the interplay between various levels at which social relations take place – including hitherto neglected aspects of health and social care experience such as inequality, body and identity, technologies of power and subjectivity and the risk society.

Ironically, the solutions to the 'problems of health and social care' are tractable to disciplines such as health and social care, because they seem to promise answers to health and social care prejudice and marginalisation (Chudacoff, 1989). For example, biomedical 'solutions' address fears about mental and physical incapacity (Longino & Powell, 2004). Medicine, with its focus on individual organic pathology and interventions, has also become a powerful and pervasive force in the definition and treatment of health and social care. The resulting 'biomedicalisation of health and social care' (Estes & Binney, 1989) socially constructs care health and social care as a process of decremental physical decline and places health and social care under the domain and control of biomedicine. It also encourages health and social cares certain forms of the politics of health and social care: a focus on health and social care as a question of social welfare, and a particular interpretation of the effects of risk and individualisation. Theories of health and social care, albeit in contrasting ways, see these phenomena as indicating particular sites of resistance in which dominant biomedical conceptualisations of health and social care are to be contested and alternative explanations can be intimated.

There has been an unprecedented rise and consolidation of theoretical publications relating to health and social care that has cut right across and through social and human sciences (Bury, 1995). Health and social care is multidisciplinary and is the principal instrument of orthodox theorising about care health and social care particularly in US, UK and Australasian academies (Phillipson, 1998; Biggs & Powell, 2001). Turner (1989) and Phillipson (1998) both acknowledge that social theory *must* be brought into the frame of analysing the people and populations it affects such as career people. This chapter faces up to this challenge and is concerned with highlighting the major theoretical ideas which have informed social understanding of health and social care in recent years. The analysis of the major theoretical ideas which have influenced our understanding of health and social care comes from mainstream theories in recent years: functionalism, Marxism, feminism and postmodernism. It is important to illuminate the contrasting theories of health and social care and health and social care, as research chapters in mainstream social health and social care have not attempted to review these theories in an attempt to understand the philosophical dimensions of human experience.

Social Theory and the Rise of Functionalist Accounts of Health and Social Care

The broad pedigree of social theories of health and social care can be located to the early post-war years with the concern about the consequences of demographic change and the potential shortage of health and social care of 'younger' workers in the United States and United Kingdom. Social health and social care emerged as a field of study which attempted to respond to the social policy implications of demographic change (Vincent, 1996). Such disciplines were shaped by significant external forces: firstly, by state intervention to achieve specific outcomes in health and social policy, and secondly, by a political and economic environment which viewed a health and social care population as creating a 'social problem' for society (Jones, 1983). This impinged mainly upon the creation of functionalist accounts of health and social care and health and social care primarily in US academies. Functionalist sociology dominated the sociological landscape in the United States from the 1930s up until 1960s (Blaikie, 1999). Talcott Parsons was a key exponent of general functionalist thought and argued that society needed certain functions in order to maintain its well-being: the stability of the family and circulation of elites in education drawing from a 'pool of talent' (Giddens, 1993). Society was seen as akin to a biological organism – all the parts (education/family/religion/government) in the system working together in order for society to function with equilibrium (Giddens, 1993).

The important point to note is that theories often mirror the norms and values of their creators and their social times, reflecting culturally dominant views of what should be the appropriate way to analyse social phenomena. The two theories which dominated American health and social care in the 1950s of health and social care and activity theory follow this normative pattern. Both disengagement and activity theories not only postulate how individual behaviour changes with health and social care but also imply how it should change:

> withdrawal may be accompanied from the outset by an increased preoccupation with himself: certain institutions may make it easy for him. (Cumming & Henry, 1961, 14)

For this variant of functionalism, this process benefits society, since it means that the death of individual society members does not prevent the ongoing functioning of the social system. Cumming and Henry (1961) further propose that the process of disengagement is inevitable, rewarding and a universal process of mutual withdrawal of the individual and society from each other with advancing health and social care – was normal and to be expected. This theory argued that it was beneficial for both the health and social care individual and society that such inequalities take place in order to minimise the social disruption caused at a person's eventual death (Neugarten, 1996).

Retirement is a good illustration of health and social care process, enabling the health and social care person to be freed of the responsibilities of an occupation and to pursue other roles not necessarily aligned to full-pay of economic generation. Cumming and Henry argued that society anticipated the loss of health and social care people through death and brought 'new blood' into full participation within the social world (cited in Katz (1996)). Bromley (1966, 136) further portends 'in care health and social care, the individual is normally disengaged and socially cared from the main streams of economic and community activity'. Not surprisingly for Bromley (1966 quoted in Bond & Coleman, 1993, 44): 'The disengagement process is graded to suit the declining biological and psychological capacities of the individual and the needs of society'.

A number of critiques exist: firstly, this theory condones indifference towards 'care health and social care' and social problems (Bond & Coleman, 1993). Both advocate a retired care health and social care as a 'natural' period of transition. In order to legitimise its generalisations, disengagement theory self-praised itself to objective and value-free rigour of research methods: survey and questionnaire methods of gerontological inquiry. In a sense, by arguing for withdrawal from work roles under the guise of objectivity is a very powerful argument for governments to legitimise boundaries of who can work and who cannot based on health and social care (Powell, 1999).

Activity theory is a counterpoint to such a fatalistic theory, since it claims a successful 'care health and social care' can be achieved by maintaining roles and relationships. Activity theory actually pre-dates disengagement theory. In the 1950s, Havighurst and Albrecht (1953 cited in Katz (1996)) insisted health and social care can be lively and creative experience. Any loss of roles, activities or relationships within care health and social care should be replaced by new roles or activities to ensure happiness, value consensus and well-being. For activity theorists, health and social care is not a natural process as advocated by Cumming and Henry.

Nevertheless, activity theory neglects issues of power, inequality and conflict between health and social care groups. An apparent 'value consensus' may reflect the interests of powerful and dominant groups within society who find it health and social care to have health and social care power relations organised in such a way. While Phillipson (1998) sees such functionalist schools as important in shaping social theory responses to them, such functionalist theories 'impose' a sense of causality on health and social care by implying you will either 'disengage' or will be 'active'. This can be argued to be a form of 'academic imperialism' where the activities of health and social care people are dictated to and from theoretical models which reconstruct health and social care and health and social care along lines of *enforced* experiences. They are very macro-orientated and fail to resolve tensions within health and social care group relations which impinge upon the interconnection of 'race', class and gender with health and social care.

Political Economy

As an intellectual backdrop against such functionalist theoretical dominance, political economy of health and social care emerged as a fashionable theory in both sides of the Atlantic, drawing from Marxian insights in analysing the capitalist complexity of modern society and how care health and social care was socially constructed to foster the needs of the economy (Estes, 1979). This critical branch of Marxist health and social care grew as a direct response to the hegemonic dominance of structural functionalism in the form of normative theory, the biomedical paradigm and world economic crises of the 1970s. As Phillipson (1998) points out in the United Kingdom, huge forms of social expenditure were allocated to career people. Consequently, not only were career people viewed in medical terms but in resource terms by governments. This brought a new perception to attitudes to health and social care and health and social care. As Phillipson (1998, 17) teases out:

> Older people came to be viewed as a burden on western economies, with demographic change… seen as creating intolerable pressures on public expenditure.

A major concern of 'political economy of care health and social care' was to challenge both the theoretical dominance of functionalist thought and biomedical models of health and social care and health and social care. The political economy approach wanted to have an understanding of the character and significance of variations in the treatment of the health and social cared and to relate these to polity, economy and society in advanced capitalist society.

The major focus is an interpretation of the relationship between health and social care and the economic structure. In the United States, political economy theory was pioneered via the work of Estes (1979) and Estes et al. (1982). Similarly, in the United Kingdom, the work of Walker (1981), Townsend (1981) and Phillipson (1982) added a critical sociological dimension to understanding health and social care in advanced capitalist societies. For Estes et al. (1982) in the United States, the class structure is perceived as the major determinant of the socio-economic position of career people in advanced capitalist society. For Estes (1979), political economy challenges the ideology of career people as belonging to a homogenous group unaffected by dominant structures in society. Estes (1979) claims political economy focuses upon an analysis of the state in contemporary societal formations. Here, we can see how Marxism is interconnected to this theory. Estes looks at how the state decides and dictates who is allocated resources and who is not. This impinges upon retirement and subsequent pension schemes. As Phillipson (1982) points out, the retirement experience is linked to the timing of economic reduction of health and social cares and enforced withdrawal from work has made many career people in the United Kingdom in a financially insecure position. Hence, the state can make and break the fortunes of its populace. Consequently, current governmental discourses of cutting public expenditure on pensions and increasingly calling for private provision legitimise ideological mystification stereotypes of 'burden' groups and populations. In the United States, Estes et al. (1982) claim that the state is using its power to *transfer* responsibility of welfare provision from the state to individuals. Indeed, blaming people for non-provision of own savings obscures and mystifies that real economic problems derive from the capitalist mode of production and political decisions (Powell, 1999).

American political economy then is a 'grand' theory drawing from Marxian historiography, locates the determining explanatory factors in the structure of society and focuses upon welfare and its contribution to the institutional decommodification of retired career people. Negative attitudes towards career people and impoverished position are best explained by the latter's loss of social worth brought about by their loss of a productive role in American society that puts premium on production (Estes et al., 1982).

Similarly, this is an argument reiterated by critical gerontological writers in the United Kingdom on the social position of career people. In particular, Townsend

(1981) observes that society creates the social problems of care health and social care through 'structured dependency' embedded in institutional agism through lack of material resources via poverty, retirement policies, negative consequences of residential care and passive forms of community care services. Townsend focuses on a 'structural' perspective of 'rules and resources' governing career people in advanced capitalism and wider social system. Importantly, Townsend claims his approach as:

> one whereby society is held to create the framework of institutions and rules within which the general problem of the elderly emerge or, indeed, "manufactured". In the everyday of the economy and the administration and development of social institutions the position of the elderly subtly changed and shaped. (Townsend, 1981, 9)

Similarly, Walker (1981) argued for a 'political economy of care health and social care' in order to understand the position of career people. In particular, Walker (1981, 77) paid attention to the 'social creation of dependency' and how social structure and relations espoused by the mode of production which helps intensify structural class marginalisation. In a similar vein, Phillipson (1982, 1986) considers how capitalism helps socially construct the social marginality of career people in key areas such as welfare delivery. The important argument to be made is that inequalities in the distribution of resources should be understood in relation to the distribution of power within society, rather than in terms of individual variation.

Feminism: Health and Social Care

Coupled with this, there has been an acceleration of feminist insights into understanding health and social care and gender as key identity variables of analysis (Arber & Ginn, 1991 and 1995). There are two important issues: first, power imbalances shape theoretical construction, and second, a group's place within the social structure influences theoretical attention they are afforded. Henceforth, because career women tend to occupy a position of lower class status, especially in terms of economic status, than men of all health and social cares and younger women, they are given less theoretical attention. According to Acker (1988 cited in Arber & Ginn, (1991)) in all known societies, the relations of distribution and production are influenced by gender and thus take on a gendered meaning. Gender relations of distribution in capitalist society are historically rooted and are transformed as the means of production change. Similarly, health and social care relations are linked to the capitalist mode of production and relations of distribution. 'Health and social care' takes on a specific meaning depending on health and social care context. For example, health and social carers work for less money than adults, who in turn work for less money than middle-health and social cared adults. Further, young children rely on personal relations with family figures such as parents. Many career people rely on resources distributed by the state.

There is a 'double standard of health and social care' with health and social care in women having particularly strong negative connotations. Career women are viewed as unworthy of respect or consideration (Arber & Ginn, 1991).

Catherine Itzin sees the double standard of health and social care as arising from the sets of conventional expectations as to health and social care-pertinent attitudes and roles for each sex which apply in patriarchal society. These are defined by Itzin as a male and a female 'chronology,' socially defined and sanctioned so that the experience of prescribed roles is sanctioned by disapproval. Male chronology hinges on employment, but a woman's health and social care status is defined in terms of events in the reproductive cycle.

Acker (1988) claim because women's value is exercised, the awareness of a loss of a youthful appearance brings social devaluation; vulnerability to pressure is penetrated by cosmeticisation. Daly (cited in Arber & Ginn, (1991)) draws a mirror health and social care between Western cosmetic surgery and the genital mutilation carried out in some African societies: both cultured practices demonstrate the pressure on women to comply with male standards of desirability and the extent of male domination. For career black women, the ideal of 'beauty' portrayed by white male culture was doubly distant and alienating, until growing black consciousness subverted health and social care language and social care and argued 'black is beautiful'.

Arber and Ginn (1991) claim patriarchal society exercises power through the chronologies of employment and reproduction, and through the sexualised promotion of a 'youthful' appearance in women. As a result, many career women suffer from a 'double jeopardy' thesis through health and social care and sexual discrimination.

Postmodern Health and Social Care

In addition to these broad and macro-based theories, there has been a vast interest in postmodern perspectives of health and social care and health and social care identity underpinned by discourses of 'better lifestyles' and increased leisure opportunities for career people due to healthier lifestyles and increased use of bio-technologies to facilitate the longevity of human experiences (Blaikie, 1999; Featherstone & Hepworth, 1993; Featherstone & Wernick, 1995; Powell & Biggs, 2002). The intellectual roots of 'postmodern health and social care' derive from Jaber F. Gubrium's (1975) sociological analysis of the discovery and conceptual elaboration of Alzheimer's disease in the United States and the establishment of boundaries between 'normal' and pathological health and social care, care health and social care are seen as a 'mask' which conceals the essential identity of the person beneath. The view of the health and social care process as a mask/disguise concealing the essentially youthful self beneath is one which appears to be a popular argument (Featherstone & Hepworth, 1989, 1993). When asked at the health and social care of 79 to describe what it felt like to be cared, the author J.B. Priestley replied:

> It is as though, walking down Shaftesbury Avenue as a fairly young man, I was suddenly kidnapped, rushed into a theatre and made to don the grey hair, the wrinkles and the other attributes of health and social care, then wheeled on health and social care. Behind the appearance of health and social care I am the same person, with the same thoughts, as when I was younger. (Pruner, 1978, 7)

There are two underlying issues for Featherstone and Hepworth (1993) which should be understood as the basis for understanding postmodern health and social care. Firstly, the health and social care of the mask alerts social gerontologists to the possibility that a tension exists between the *external* appearance of the body and face and functional capacities and the *internal* or subjective sense of experience of personal identity which is likely to become prominent as health and social care traverses through the life-course.

Secondly, career people are usually 'fixed' to roles without resources which does not do justice to the richness of their individual experiences and multi-facets of their personalities. Idealistically, Featherstone and Hepworth argue that a postmodern perspective would deconstruct such realities, and health and social care should be viewed as fluid with possibilities not constrained by medical model decline discourses.

According to Powell and Biggs (2000), the direct use of new technologies to modify either the appearance or performance of health and social care identity is symptomatic of postmodern times. To paraphrase Morris (1998), technologies here give out the promise of 'utopian bodies'. Indeed, Haraway's (1991) (cited in Powell and Biggs (2002)) original reference to cyborgic fusion of biological and machine entities has been enthusiastically taken up by postmodern health and social care. The list of technologies available extends beyond traditional prosthesis to include virtual identities created by and reflected in the growing number of 'silver surfers' using the Internet as a free-floating form of identity health and social care. Thus, Featherstone and Wernick (1995, 3) trill that it is now possible to 'Re-code the body itself "as biomedical and information technologies make available" the capacity to alter not just the meaning, but the very material infrastructure of the body. Bodies can be re-shaped, remade, fused with machines, empowered through technological devices and extensions'.

Coupled with this, the control of the health and social care body had been enhanced by external constraining virtue of the corset, and contemporary shaping has involved active working, through exercise and diet. The multiplication of magazine chapters, self-help manuals and diet and exercise clubs, extending through midlife and beyond, also bear witness to the popularity of attempts to work on the self in this way.

The use of diet and exercise as techniques specifically related to later adulthood is closely related to the growth of leisure and a lifestyle approach to the creation of late life identities (Turner, 1989; Powell & Biggs, 2000, 2002). It therefore resonates beyond the simple fuelling and repair of the bodily machine to include a continual re-creation of the self within a particular social discourse. This discourse closely associates the construction of a healthy lifestyle with positive self-identity.

Indeed, closely related to postmodern health and social care is also a small body of knowledge pertaining to 'Foucauldian health and social care' deriving in Canadian (Katz, 1996; Frank, 1998) and UK (Biggs & Powell, 1999; Powell & Biggs, 2000, 2002; Powell & Cook, 2000; Wahidin & Powell, 2001) academies. This Foucauldian theoretical development attempts to understand health and social care and health and social care through conceptual exploration of power/knowledge and how surveillance practices from professionals such as 'medics' or 'social workers' further marginalise, normalise and provide shape to the experiences of career people (Powell & Biggs, 2000).

Conclusion

These theories have been at the forefront of understanding care health and social care in occidental academies. Taken together, these theoretical currents have been influential in providing social health and social care with a rich social dimension. Such social theories have been used also to analyse pressing social issues such as elder abuse, the gendered nature of health and social care, the politics of power relations between career people and state/society and community care. The purpose of this chapter has been to amalgamate the key ideas of social theories of health and social care in order to stress the importance of social philosophy to understanding health and social care. While these mainstream theories have dominated an understanding of the what and the why of health and social care inequities, there have been more evocative theories that have emanated that provides questions of power, surveillance and resistance: the theory of Michel Foucault. The next chapter provides an in-depth exploration of his work before we can see the visibility of his conceptual tools to open up health and social care policy, practice a new perspectives.

Chapter 2
New Perspectives: Foucault and Applicability to Health and Social Care

Introduction

In order to situate Michel Foucault's pathbreaking thinking in relation to health and social care, we need to understand the background against which he formulated his views. If we return to an earlier discussion on the intellectual climate in post-1945 France, two views emerge in relation to Husserlian phenomenology. First, structuralism, and second, a hermeneutics which did not resort to the transcendental subject, but argued that the source of meaning production should be sought in the social practices and texts which are fundamental to human activity. For the purposes of this section, we may trace this latter intellectual development relevant to health and social care from Martin Heidegger. In the process, we shall be required to consider the generation of meaning without reference to an autonomous subject and to utilise the idea of discourses without allusion to underlying rules of structure. In other words, we have to move beyond the limitations of structuralism and hermeneutics in order to develop an 'interpretive analytics' (Dreyfus & Rabinow, 1982a, b; May & Powell, 2007).

The fusion of these different ideas in Foucault's work results in an approach to the history of thought that is freed: 'from its subjection to transcendence' (1989, 203). In this journey, we are also required to take two further steps. First, the relationship between the subject and truth should be viewed as an effect of knowledge itself. As Foucault put it:

> what if understanding the relation of the subject to the truth, were just an effect of knowledge? What if understanding were a complex, multiple, non-individual formation, not 'subjected to the subject', which produced effects of truth?(Foucault in Elders, 1974, 149)

Knowledge should not, therefore, be viewed as 'theoretical' separate from the realm of 'practice'. Instead, knowledge becomes a practice which has the effect of constituting particular objects – non-theoretical elements – which are part of practice itself. Similarly, the theory of knowledge and the subject of knowledge become

J. Powell, *New Perspectives on Health and Social Care*, International Perspectives on Social Policy, Administration, and Practice,
https://doi.org/10.1007/978-3-031-25432-1_2

fused, and the subject and truth are not related in the ways suggested earlier, but are seen as part of a relationship between knowledge and power that is socially constructed (Gutting, 2003):

> The important thing here, I believe, is that truth isn't outside power, or lacking in power: contrary to a myth whose history and functions would repay further study, truth isn't the reward of free spirits, the child of protracted solitude, nor the privilege of those who have succeeded in liberating themselves. Truth is a thing of this world: it is produced only by virtue of multiple forms of constraint. And it induces regular effects of power. Each society has its of truth, its 'general politics' of truth. (Foucault, 1980a, b, 131)

What Foucault is doing here is rejecting certainty in social and political life and arguing that there is no universal understanding beyond history and society. In the process, a 'history of the present' is written, according to which history is not excavated to reveal a deeper meaning but, instead, is viewed as a 'battleground' between relations of power. Now this begins, albeit in a different guise, to look like structuralism: that is, to view events as manifestations of a deeper, underlying reality. How is this issue tackled in his work?

In commentating on those who have labelled him a structuralist, Foucault writes: 'I have been unable to get it into their tiny minds that I have used none of the methods, concepts, or key terms that characterise structural analysis' (1989, xiv). Fine! Beyond assertion, however, what are his justifications for this position? After all, the opening for the type of analysis we have considered so far occurred in Althusserian structuralism. Ideologies are seen as ways of constructing subjects and their perspectives on the world, all of which takes place against the backdrop of a Marxist model of social reality.

The first point we might observe here is that Foucault eschews grand theoretical projects of the Althusserian type. Perhaps the most simple way to view his relationship with structuralism is by examining his analysis of historical 'events'. What we find here is a refusal to read historical 'events' as manifestations of deeper social structures, or as located on particular levels, accompanied by a focus upon the seemingly marginal as indicative of relations of power. Events are thereby seen to differ in their capacity to produce effects:

> The problem is at once to distinguish among events, to differentiate the networks and levels to which they belong, and to reconstitute the lines along which they are connected and engender one another. From this follows a refusal of analyses couched in terms of the symbolic field or the domain of signifying structures, and a recourse to analyses in terms of the genealogy of relations of force, strategic development, and tactics. Here I believe one's point of reference should not be to the great model of health and social care and signs, but to that of war and battle. (Foucault, 1980a, b, 114)

If Foucault does not reach for structuralism, neither does he resort to grounds in the universal preconditions of human understanding (Heidegger). What he calls genealogy, following Nietzsche, is an analysis of cultural practices without a reference point outside of history itself. The motivating question underlying this approach is 'What are we today?' (1988, 145). For it is history which makes us what we are and what we think of ourselves as being. This is a very different starting point from, say, an analysis of the limits which are continually placed upon what is seen as our 'hidden' creativity as found in a critical social theory based on philosophical anthropology (Honneth & Joas, 1988). The manifestations of such a project we have covered in relation to a 'dialectic of freedom' whose aim is to determine the good and bad elements in the Enlightenment. Genealogy, in contrast, does not involve a search for any 'origin':

> On the contrary, it will cultivate the details and accidents that accompany every beginning; it will be scrupulously attentive to their petty malice; it will await their emergence, once unmasked, as the face of the other. The genealogist needs history to dispel the chimeras of the origin. (Foucault, 1984a, b, c, 80)

We have come across other universal grounds for analysis through the employment of modified forms of reason, none more powerfully argued than by Jürgen Habermas. Yet the issues motivated by such an approach are still couched in terms of the ability to provide grounds for distinguishing between the true and the false, not in terms of the formation of truth itself. It is via this focus in Foucault's work that we uncover the legacy of Nietzsche.

Unlike Kant, Nietzsche did not ask what the grounds or basis for our knowledge was, but why knowledge was necessary in the first place. He argued that it was the belief we have in knowledge that is of importance, not whether it is true or false. This belief is seen to be indicative of a 'will to truth' which, as part of the 'will to power', involves the desire to affirm life as it appears to us. Truth now becomes inverted to be a fiction that is 'invented':

> What urges you on and arouses your ardour, you wisest of men, do you call it 'will to truth'? Will to the conceivability of all being: that is what I call your will! You first want to make all being conceivable: for, with a healthy mistrust, you doubt whether it is in fact conceivable. But it must bend and accommodate itself to you! Thus will your will have it. It must become smooth and subject to the mind as the mind's mirror and reflection. This is your entire will, you wisest men; it is a will to power; and that is so even when you talk of good and evil and of the assessment of values. (Nietzsche in Hollingdale, 1977, 224)

With the Nietzschean turn in place, the object of Foucault's analyses became the production of discourses that provide justifications for actions: 'my problem is to see how men govern (themselves and others) by the production of truth' (1991c, 79). With a focus upon how the truth is constructed in a manner that permits, from the point of view of practice, a distinction to be made between the true and the false, Foucault's use of language as an analytic resource becomes distinct from any we have come across before.

Thus far, we have considered the role of Foucault in social life in a number of ways: De Saussure's structuralism, as a 'form of life' (Wittgenstein) analysed in terms of speech acts (Austin, Searle), a fusion of speech act theory onto the problematic of rationalisation (Habermas) and the relationship between utterance and institutional authority (Bourdieu). We could undertake genealogical analysis by, for example, examining the rules of governing an utterance and then derive a set of rules which produced the utterance in the first place. Our results may then be transposed from context to context. Once again, however, Foucault throws a spanner into the works of these approaches.

Foucault's analysis of discourse does not contain the idea of knowledge production via a meaning-producing subject, or rules that govern speech acts, nor an allusion to rationality that poses some absolute value in reason (1991c). Given this, his approach cannot be situated within any of the above positions. Nevertheless, he sees discourses as a unity of statements in what is termed a 'discursive formation'. The question is what provides their unity? Foucault rejects four possible answers to this question. First, that there is a well-defined field of objects to which they refer; second, that they possess a clear normative basis; third, that they possess what he calls an 'alphabet of notions'; and finally, that they are characterised by 'the permanence of a thematic' (1989, 37).

With these possibilities rejected, what we uncover is the existence of possibilities and differences in structures and rules. Where it does become possible to describe these 'systems of dispersion' in terms of objects, statements, concepts or themes that make up a 'regularity', then we may say:

> for the sake of convenience, that we are dealing with a discursive formation ... The conditions to which the elements of this division (objects, mode of statement, concepts, thematic choices) are subjected we shall call the rules of formation. The rules of formation are conditions of existence (but also of coexistence, maintenance, modification and disappearance) in a given discursive formation. (Foucault, 1989, 38)

In a description of the events of discourse, the question then becomes 'How is it that one particular statement appeared rather than another?' (1989, 27). However, there exists the potential for terminological confusion in Foucault's idea of 'statements'. Dreyfus and Rabinow thus propose the term 'serious speech acts' (1982a, b, 48). This neatly encapsulates Foucault's adherence to the study of relations between speech acts and their continual formation and transformation where they are validated by particular procedures and the increasing army of 'experts' which characterise modern societies. It also takes account of Foucault's acceptance of the idea we have come across before to the effect that speech acts are formulated against a background of everyday assumptions, while the idea of 'serious' retains his emphasis upon the generation of meaning within discourses.

With these themes and arguments in mind, we can now say that Foucault did not resort to a depth-hermeneutics whereby, via an excavation of the background assumptions which inform everyday interpretations, a 'truth' is revealed to exist. Further, that he was not content to view meaning as ultimately residing in a transcendental subject, nor to eliminate meaning through the employment of structuralist insights (Layder, 2006). His concerns were to empirically examine the effects of social practices without recourse to some idea of their intrinsic homogeneity. It is to an understanding of this project in relation to power and the subject that we now turn.

Power and the Modern Subject

The above situates Foucault's approach to analysing social phenomena in terms of the politics of truth, but it says little of the effects of these discursive regimes. For this purpose, Foucault turned towards analyses of the exercise of power in contrast to a consideration of the forms of knowledge which create a sense of social order. In more specific terms, his objective became 'to create a history of the different modes by which, in our culture, human beings are made subjects' (1982, 208).

To return to our earlier discussion. The position which the analyst occupies in this process of inquiry is still not that of the 'bearer of universal values' (1980a, b, 132). It is, however, as someone who is situated within society and does not speak for the truth, but in a 'battle' regarding its status and the political and economic role that it plays in social relations (1980a, b, 132). In other words, what is taken as being self-evident in the present is viewed as part of an historical process. From this point of view, Foucault's (1977a, b, c) study on imprisonment and punishment aimed:

> to write a history not of the prison as an institution, but of the practice of imprisonment: to show its origin or, more exactly, to show how this way of doing things - ancient enough in itself - was capable of being accepted at a certain moment as a principal component of the penal system, thus coming to seem an altogether natural self-evident and indispensable part of it. (Foucault, 1991a, b, c, d, 75)

Three lines of inquiry are required in order to understand how the subject is objectified. First, how particular sciences contribute to this process. Second, a study of the 'practices' that divide people from others, as well as within themselves: for example, the mad and sane, the sick and healthy and the criminal and non-criminal. Third, in what ways people then turn themselves into subjects (Foucault, 1982, 208). Just to reiterate: the form of this analysis is opposed to the search for origins and ideal significations, as well as the notion that we possess an intrinsically rational and coherent sense of self.

At this point, we find another central aspect to Foucault's work which causes interpretative confusion. Quite simply, it is the tension which exists between his approach to the study of social relations and a deep-seated idea within Western cultures: that is, the belief that the condition of freedom is the absence of power and thus power is always repressive. To commit oneself to this tradition would be to fall into the trap that Foucault's inspired genealogies seek to avoid. For instance, when it comes to the individual in relation to the means of production, or as involved in relations of signification, we turn to economic history, linguistics and semiotics to shed light on these matters. When it comes to power all we ever ask is 'what legitimates power?' (1982, 209). However, with the individual implicated in these relations, this approach cannot suffice. In order to understand how the subject is objectified, this idea of power needs total revision. This is where Foucault's 're-conceptualisation' of power is of central importance, and we need to understand what he means by this before we continue with our discussion.

Power is not inevitably vested in the state, nor localised in, say, relations between the citizen and state. Instead, the effects of the domination of power are attributable to various 'dispositions, manoeuvres, tactics, techniques, functionings' (1991d, 26). Its properties are invested in those over whom it is exercised, so it not only exerts pressures upon people but is also transmitted by them: 'Power comes from below; that is, there is no binary and all-encompassing opposition between rulers and ruled' (1979, 94). Power is part of the social body as a whole – in the machinery of production, families, limited groups and institutions (1979, 94). Thus, power is not a property of capitalism, patriarchy or the bourgeoisie – it is a strategy that has evolved as part of the characteristics of modern society. Nor should power be identified with an individual who possesses or exercises it by right of birth:

> Not the domination of the King in his central position, therefore, but that of his subjects in their mutual relations: not the uniform edifice of sovereignty, but the multiple forms of subjugation that have a place and function within the social organism. (Foucault in Kelly, 1994, 34)

Given this, if we conceive of power as emanating from one source, we will not understand its effects. When it comes to politics and power, the sovereign's head must be cut off. Further, to think of power as repressive (the repressive hypothesis) is to neglect its positive aspects. In this sense, power becomes 'a machine in which everyone is caught' (Foucault, 1980a, b, 156).

Yet a question still remains: if power is transmitted in this way, where does it come from in the first place? This is one of the most difficult concepts to grasp. After all, if knowledge and power are linked, one thinks of Marx and the idea that those who are in charge of the means of production are, at the same time, in charge of the means of mental production. An illuminating discussion of this issue can be found in an interview with Foucault, conducted by Jean-Pierre Barou and Michelle Perrot (Foucault, 1980a, b, 156).

Let us take the example of a factory: here, relations of power may be seen in terms of how 'individuals try to conduct, to determine the behaviour of others' (1991b, 18). The system of power is pyramidal and it occurs between the health and social carers and the health and social cared. However, the strategic apex (the health and social carers) is not the source of power, nor is a principle or goal which organisational members are expected to invest with legitimacy in terms of their practices. Power and its techniques are used throughout the factory. Therefore, it cannot be simply identified with a particular group:

> These tactics were invented and organised from the starting points of local conditions and particular needs. They took shape in piecemeal fashion, prior to any class strategy designed to weld them into vast, coherent ensembles. It should also be noted that these ensembles don't consist in a homogenisation, but rather a complex play of supports in mutual eng-health and social care, different mechanisms of power which retain all their specific character. (Foucault, 1980a, b, 159)

We can draw three points from this quote. First, in his analysis of power, Foucault is concentrating on the local and contingent. Therefore, as noted before, he is not prepared to 'read off' practices against some universal idea of rationality. Rather, he is concerned to examine the ways in which particular rationalities are deployed, leaving the connections to be established among them open to investigation (Dean, 2007). Second, power should not be seen, from this point of view, as existing above, but alongside developments in the forces and relations of production. Power is transformed along with, but not by these changes and may be found in both socialist and capitalist societies (Foucault, 1980a, b, 160). Third, the workings of this form of power can be traced far back into history.

In terms of tracing the history of this form of power, it was during the seventeenth, eighteenth and nineteenth centuries that a new form of political rationality found its target. When faced with the issue of population growth, political economists examined it in terms of wealth accumulation and the productive capabilities of labour, with the causes of poverty being conceived of in terms of disease, idleness, etc. The social reformer and philosopher Jeremy Bentham (1748–1832), on the other hand, 'poses the question in terms of power - population as object of relations of domination' (Foucault, 1980a, b, 151). It was the economic changes occurring in the eighteenth century that:

> made it necessary to ensure the circulation of effects of power through progressively finer channels, gaining access to individuals themselves, to their bodies, their gestures and all their daily actions. By such means power, even when faced with ruling a multiplicity of men, could be as efficacious as if it were being exercised over a single one. (Foucault, 1980a, b, 151–152)

The differentiation of the population thus required this new form of power. The simple domination of one group by another was no longer a feasible technique given the increasing complexity of society. Contrary to simple ideas of coercion, power has both a political and economic cost. If you are too violent in controlling a population, you risk revolt. If you intervene too frequently, then resistance and disobedience may result. Disciplinary technologies and regulatory procedures must inform and produce a normative basis which spreads throughout the social body while masking the operation of power that underlies this process (Katz, 2007). In this way, the modern subject becomes inextricably linked to the society of which they are a part through turning themselves into an object:

> There is no need for arms, physical violence, material constraints. Just a gaze. An inspecting gaze, a gaze which each individual under its weight will end up interiorising to the point that he is his own overseer, each individual thus exercising this surveillance over, and against, himself. A superb formula: power exercised continuously and for what turns out to be a minimal cost. (Foucault, 1980a, b, 155)

The body, space and time are routinely controlled (Turner & Dumas, 2006). What Foucault calls 'bio-power' relies upon the scientific categorisation of the population, in terms of being an object of systematic and sustained political intervention, as well as encompassing a focus upon the human body as an object of control and manipulation. Collectively, these 'technologies' centre around the 'objectification' of the body. They form a 'disciplinary technology' whose aim is to forge the individual, with 'normalisation' being one technique (Foucault, 1977a, b, c). This is best exemplified in enclosed situations, such as a prison, school, hospital or, as above, the factory, where the control of space and time is more readily apparent. It is here that we find:

> a whole micro penality of time (lateness, absences, interruptions of tasks), of activity (inattention, negligence, lack of zeal), of behaviour (impoliteness, disobedience), of speech (idle chatter, insolence), of the body ('incorrect' attitudes, irregular gestures, lack of cleanliness), of sexuality (impurity, indecency). (Foucault, 1991d, 178)

This disciplinary technology permits supervision and control with interiorisation being the most effective deployment of power. In this respect, Foucault suggests that there are three ways in which 'self-examination' has developed over time. First, the Cartesian form such that thoughts are considered in relation to their correspondence with reality. Second, the ways in which our thoughts correspond with, or relate to, rules. Third, the relationship between a thought which is 'hidden' and some inner impurity:

> At this moment begins the Christian hermeneutics of the self with its deciphering of inner thoughts. It implies that there is something hidden in ourselves and that we are always in a self-illusion which hides the secret. (Foucault, 1988, 46)

The discourse of constituting the subject as object thereby also requires the discourse of the subject, and this too has changed over the course of history. The means of exposing the truth of ourselves in Christianity involved being subjugated to a master. This model of obedience means that once the confession was verbalised, it was accompanied by a renunciation of the sense of self and 'will' as part of a continual process of self-renunciation: that is, disclosing oneself to an authority. This is a 'permanent verbalisation'. However, what then occurs is that the interpretative sciences come to play this central role. After all, it is they who 'claim to be able to reveal the truth about our psyches, our culture, our society -truths that can only be understood by expert practitioners' (Dreyfus & Rabinow, 1982a, b, 180):

> From the eighteenth century to the present, the techniques of verbalization have been reinserted in a different context by the so-called human sciences in order to use them without renunciation of the self but to constitute, positively, a new self. To use these techniques without renouncing oneself constitutes a decisive break. (Foucault, 1988, 49)

This is accompanied by the state performing 'a modern matrix of individualisation' (1982, 215). Together, these are the new forms of 'pastoral power'. Nevertheless, while claiming to speak the truth, these interpretative sciences are actually part of the rationalities which inform a political technology that saturates everyday life (Dean, 2007). Importantly, however, this should not be conceived of as being a deterministic relationship.

To illuminate the indeterminacy of these practices and hence the possibilities for their transformation, the focus of Foucault's investigations changes. He now considers not so much the ways in which we are constituted as both subjects and objects, but the effects of this on our relations with others and the implications for the relationship between power, truth and the self. Now these are not discrete topics, but ultimately complementary strategies of historical investigation (Foucault, 1988, 15). The body then becomes not only the object or target of power, but also the centre for resistance and opposition. After all, contemporary society is increasingly characterised by struggles against the exercise of power:

> opposition to the power of men over women, of parents over children, of psychiatry over the mentally ill, of medicine over the population, of administration over the ways people live. (Foucault, 1982, 211)

Foucault asks the empirical question 'what happens' when power is exercised? As opposed to becoming involved in endless debates over what power 'is' and where it comes from (1982, 217)? Resistance to forms of power is the starting point for this investigation. These are struggles for new subjectivities which encompass the right to be different and distinguish oneself from the ways in which the effects of particular 'technologies of the self' routinely constitute people: 'These struggles are not exactly for or against the "individual", but rather they are struggles against the "government of individualisation"' (1982, 212). This makes sense if we remember that the individual is not constituted before the effects of power because:

> it is already one of the prime effects of power that certain bodies, certain gestures, certain discourses, certain desires, come to be identified and constituted as individuals. (Foucault, 1980a, b, 98)

It is through 'historical investigation' that scholars can understand the present. However, when utilising historical inquiry, scholars should 'use it, to deform it, to make it groan and protest' (Foucault, 1980a, b, 54). Foucault (1972, 1977a, b, c) uses his methodological 'tools' to disrupt history at the same time as giving history a power/knowledge re-configuration that makes his approach so distinctive and relevant to social theory and cultural analysis. In *The Archaeology of Knowledge*, Foucault (1972) discusses 'archaeology' as the analysis of a statement as it occurs in the historical archive. Further, archaeology 'describes discourses as practices specified in the element of the archive' (1972, 131), the archive being 'the general system of the formation and transformation of statements' (1972, 130). While an understanding of the archive would ask what rules have provided for a particular statement, the analysis of discourse asks a different question: 'how is it that one particular statement appeared rather than another?' (1972, 27).

The use of an archaeological method explores the networks of what is said and what can be seen in a set of social arrangements: in the conduct of an archaeology, there is a visibility in 'opening up' statements. Archaeology charts the relationship between statements and the visible and those 'institutions' which acquire authority and provide limits within which discursive objects may exist.

In this approach, we can see that the attempt to understand the relations between statements and visibility focuses on those set of statements that make up institutions such as prisons – instructions to prison officers, statements about time-tabling of activities for inmates and the structure and space of the carceral institution itself. This leads to the production of 'a whole micro - penality of time (lateness, absences, interruptions of tasks), of activity (inattention, negligence, lack of zeal), of behaviour (impoliteness, disobedience), of speech (idle chatter, insolence), of the body (incorrect attitudes, irregular gestures, lack of cleanliness), of sexuality (impurity, indecency)' (Foucault, 1977a, b, c, 178).

The crucial point is that this approach draws our attention to the dynamic interrelationship between statements and institutions. Secondly, the attempt to describe 'institutions' which acquire authority and provide limits within which discursive objects may act focuses again on the institution which delimits the range of activities of discursive objects (Powell & Biggs, 2000) – it is at this point that an exploration of the architectural features of the institution would be used to understand spatial arrangements. In a similar context, Goffman (1968) wrote about how spatial arrangements of 'total institutions' operate to provide care and rehabilitation at an official level and capacity, underneath the surface. Such institutions curtail the rights of those within them: 'Many total institutions, most of the time, seem to function merely as storhealth and social care dumps for inmates ... but they usually present themselves to the public as rational organizations designed consciously, through and through, as effective machines for producing a few officially avowed and officially approved ends' (Goffman, 1968, 73).

A key difference between Goffman and Foucault's interpretations of institutions would be, however, that whereas Goffman sees total institutions as an aberration, untypical of society as a whole, Foucault's critique assumes that the carceral element of institutional life encapsulates a core feature of social life. A reason for wanting to study prisons, aside from its prior neglect, was 'the idea of reactivating the project of a "genealogy of morals", one which worked by tracing the lines of what one might call "moral technologies". In order to get a better understanding of what is punished and why, I wanted to ask the question: how does one punish?' (Foucault, 1989, 276).

Foucault never felt totally comfortable with archaeological analysis and felt that discourses did not reveal the irregularities between ongoing within social practices. As a result, he developed his methodology during the course of his investigations.

Genealogy

Foucault acquired the concept of 'genealogy' from the writings of Nietzsche. Genealogy still maintains elements of archaeology including the analysis of statements in the 'archive' (Foucault, 1977a, b, c, 1980a, b, 1982). With genealogy, Foucault (1977a, b, c) added a concern with the analysis of power/knowledge which manifests itself in the 'history of the present'. As Rose (1984) points out, genealogy concerns itself with disreputable origins and 'unpalatable functions'. This can, for example, be seen in relation to psycho-casework, care health and social care and probation practice (Biggs & Powell, 1999, 2001; May, 1991, 1994). As Foucault found in his exploration of psychiatric power: 'Couldn't the interweaving effects of power and knowledge be grasped with greater certainty in the case of a science as "dubious" as psychiatry?' (1980a, b, 109).

Genealogy also establishes itself from archaeology in it approach to discourse. Whereas archaeology provides a snapshot, a 'slice' through the discursive nexus, genealogy focuses on the processual aspects of the web of discourse – its ongoing character (Foucault, 1980a, b). Foucault did attempt to make the difference between them explicit: 'If we were to characterise it in two terms, then "archaeology" would be the appropriate methodology of this analysis of local discursiveness, and "genealogy" would be the tactics whereby, on the basis of the descriptions of these local discursivities, the subjected knowledge's which were thus released would be brought into play' (Foucault, 1980a, b, 85).

Foucault is claiming that archaeology is a systematic method of investigating official statements such as dispositifs (McNay, 1994). Genealogy is a way of putting archaeology to practical effect, a way of linking it to cultural concerns: 'A genealogy of values, morality, asceticism, and knowledge will never confuse itself with a question for their "origins", will never neglect as inaccessible the vicissitudes of history. On the contrary, it will cultivate the details and accidents that accompany every beginning; it will be scrupulously attentive to their petty malice; it will await their emergence, once unmasked, as the face of the other. Wherever it is made to go, it will not be reticent – in "excavating the depths", in allowing time for these elements to escape from a labyrinth where not truth had ever detained them. The genealogist needs history to dispel the chimeras of the origin, somewhat in the manner of the pious philosopher who needs a doctor to exorcise the shadow of his soul' (Foucault, 1984a, b, c, 80).

Foucault's use of genealogy cannot be divorced from an understanding of power, nor can the constitution of the subject. With this in mind, our approach will be to consider his analytical ingenuity via an examination of different modes through which 'subjectivity' is constituted. Foucault (1982, 1983) grounded this as a pivotal mode of analysis that has been deployed in reflections on his own life (Miller, 1993). Subjectivity appears as both an experiential and discursive strategy that 'goes beyond theory' (Dreyfus & Rabinow, 1983) and provides us with a way to problematise the explanatory value and relevance of his analyses.

We will discuss Foucault's approach to subjectivity in terms of classification, dividing and self-subjectification practices. These operate in ways to structure subjectivity under the auspices of the 'rise of modernity' where, commencing in the seventeenth century, the social sciences, early capitalism and institutions began to co-ordinate new ways of objectifying 'populations' in Western societies. In Foucault's analysis, the realm of the 'social' becomes the object of inquiry. Here, the term 'social' means 'the entire range of methods which make the members of a society relatively safe from the effects of economic fluctuation by providing a certain security' (Donzelot, 1980 p. xxvi). Thus, in *Discipline and Punish*, the study 'traces the historical emergence of the social as a domain or field of inquiry and intervention, a space structured by a multiplicity of discourses emanating from the human sciences which, in their turn, are derived from, yet provide, a range of methods and techniques for regulating and ordering the social domain' (Smart, 1983).

Foucault's (1980a, b) main concern was to show that the 'truth' status of a knowledge derives from the field in which it, as a discourse, is employed and not from the interpretation of subjects' thoughts or intentions. Discourses are powerful in that they operate as a set of rules informing thought and practice, and the operation of these decides who or what is constituted as an object of knowledge. The relationship between the subject and truth should be viewed as an effect of knowledge itself. Quite simply, the subject is not the source of truth. As Foucault put it: 'what if understanding the relation of the subject to the truth, were just an effect of knowledge? What if understanding were a complex, multiple, non-individual formation, not "subjected to the subject", which produced effects of truth?' (Foucault in Elders, 1974, 149).

Knowledge is not separate from the realm of 'practice'. Knowledge is a practice that constitutes particular objects – non-theoretical elements – that are part of practice itself. Knowledge and the subject of knowledge are fused as part of the relationship between knowledge and power that is socially constructed: 'The important thing here, I believe, is that truth isn't outside power, or lacking in power: contrary to a myth whose history and functions would repay further study, truth isn't the reward of free spirits, the child of protracted solitude, nor the privilege of those who have succeeded in liberating themselves. Truth is a thing of this world: it is produced only by virtue of multiple forms of constraint. And it induces regular effects of power. Each society has its regime of truth, its "general politics" of truth' (Foucault, 1980a, b, 131).

Foucault is deliberately questioning the individual subjects' will to construct as he sets about exploring the relationship between 'discourse' and 'subjectivity'. What emerges is a grounded understanding of power/knowledge construction and reconstruction as discourses transform people into types of subjects – as classifying practices. Through these techniques of knowing, human attributes are studied, defined, organised and codified in accordance with the meta-categories of what is 'normal'. Classifying practices and techniques of normalisation designate both the objects to be known and the subjects who have the authority to speak about them. Discourses thus encompass both the objective and subjective conditions of human relations (1973, 232), and these emerging forms of social regulation, characterised by notions of discipline, surveillance and normalisation, are core to his theoretical studies (Foucault, 1977a, b, c).

The knowledge and practices are also referred to as 'epistemes' which are 'the total set of relations that unite at a given period, the discursive practices that give rise to epistemological figures, sciences and formalised systems' (Foucault, 1972, 191). Social science disciplines, in different ways, order the status of those who can validate knowledge through inquiry. Foucault designates a discourse's function of dispersing subjects and objects as its 'enunciative modality' (Foucault, 1972, 50). This modality encompasses roles and statuses and demarcated subject positions. Together they act to structure the space of regulation where the professionalisation of knowledge is instigated.

Dividing practices are deployed in order to maintain social order – to separate, categorise, normalise and institutionalise populations. In *Madness and Civilization* (1973a), *Birth of the Clinic* (1975) and *Discipline and Punish* (1977), Foucault illustrates how 'unproductive' people were identified as political problems with the 'rise of modernity'. The state divided these people into 'the mad', 'the poor' and 'the delinquent' and subsequently disciplined them in institutions: asylums, hospitals, prisons and schools (Foucault, 1977a, b, c). These exercises of disciplinary power were targeted at the subject and constituted techniques in these institutions. For instance, as we noted earlier, in *Discipline and Punish*, Foucault argues that since the eighteenth century, prison authorities increasingly employed subtle regulatory methods of examination, training, time-tabling and surveillance of conduct on offenders in which we find a whole 'micro-penality'. Overall, dividing practices are seen as integral to the rationalism of the Enlightenment narratives of liberty, individuality and rights and as fusing with governmental forms of human calculation and audit.

The previous modes of classification and dividing practices coexist. Professions examine, calculate and classify the groups that governments and institutions regulate, discipline and divide. The third mode of self-subjectification is more intangible. These practices designate the ways in which a person turns themselves into social subjects. Foucault claims that self-subjectification entails the deployment of technologies of the self: 'Techniques that permit individuals to affect, by their own means, a certain number of operations on their own bodies, their own souls, their own selves, modify themselves, and attain a certain state of perfection, happiness, purity, supernatural power' (Foucault, 1982, 10). In Foucault's work self-subjectification, practices proliferate in the domain of sexuality because the occupying sciences of medicine, psychology and psychoanalysis obligate subjects to speak about their sexuality. In turn, these sciences characterise sexual identity as esoteric and dangerous (Foucault, 1980a, b). Thus, the association of sexual truth with self-subjectification gives 'experts' their power.

Self-subjectification practices inter-relate with classification and dividing practices to construct modern subjects. For instance, subjects are created by human sciences that classify problems, identities and experiences, the systems of power that divide, stratify and institutionalise types of 'elderly' subjects and the technologies of the self that impose upon individuals the reflexive means to problematise themselves. What Foucault seems to be confronting us with is a disturbing vision that our ideas about the depth of human experience are simply cultural veneers that exist in an interplay between power and knowledge. Shumway (1989) calls this a 'strategy of exteriority': a strategy that 'does not stem from a claim that the true being plain and visible, but from a rejection of the claim that the true is systematically disguised' (1989, 26). Foucault's analysis of subjectification practices highlights techniques used by administrative powers to problematise subjects and the games of truth employed by those who seek to know them through classification techniques.

Foucault juxtaposes his axis of classifying, dividing and self-subjectification practices with one that delineates three domains of subjectivity: the body, the population and the individual. These domains elaborate how modes of subjectivity traverse modern social relations.

The Body

The 'body' is a subject of discursive and political inscription. In *Discipline and Punish*, Foucault (1977a, b, c) claims that penal practices produce the 'soul' of the offender by disciplining the body and corporealising prison spaces. In prisons, the body's most essentialist needs – food, space, exercise, sleep, privacy, light and heat – become the materials upon which schedules, curfews, time-tables and micro-punishments are enacted. The body discipline developed in prisons has parallels throughout the broader disciplinary society. Indeed, the success of modernity's domination over efficient bodies in industry, docile bodies in prisons, patient bodies in clinical research and regimented bodies in schools and residential centres attest to Foucault's thesis that the human body is a highly adaptable terminus for the circulation of power relations.

It would be a mistake to believe Foucault is alone in arguing that the rule of the body is fundamental to modern politico-economical and professional regimes of power. Critiques of the domination of the body were the mainstay of Frankfurt theorists such as Adorno and Horkheimer (1944) long before Foucault's work. As he noted of their work: 'As far as I'm concerned, I think that the Frankfurt School set problems that are still being worked on. Among others, the effects of power that are connected to a rationality that has been historically and geographically defined in the West, starting from the sixteenth century on. The West could never have attained the economic and cultural effects that are unique to it without the exercise of that specific form of rationality' (Foucault, 1991a, b, c, d, 117). Foucault's contribution, however, is to locate the ways in which 'bio-power' and disciplinary techniques construe the body as an object of knowledge. For example, *The History of Sexuality* depicts the dominion with which nineteenth-century experts constructed a hierarchy of sexualised bodies and fragmented the population into groups of 'normal', 'deviant' and 'perverted'.

While Foucault's definition of the body has inspired numerous debates, the task of refinement and problematisation have largely been the province of feminist scholars. Foucault has been criticised for his lack of sensitivity and attention to gender inequality and women's history, thereby requiring theoretical revision in order to overcome such limitations (Powell & Biggs, 2000). Feminists have stressed that the body is both a site of regulation, where gendered identities are maintained, and a site of resistance, where they are undone and challenged. McNay (1993) agrees with Foucault that 'sexuality is produced in the body in such a manner as to facilitate the regulation of social relations' (1993, 32). However, contra-Foucault, she notes that not all aspects of sexuality, corporeality and desire are products of power relations. Passionate social relationships based on friendship do not necessarily facilitate intense forms of surveillance and regulation. 'Friends' can transform disciplinary spaces and health and social care in disrupting practices. Similarly, Butler (1990, 140–141) claims that ritualised body performances that bind women to fictional feminine identities can also become deconstructive performances that expose the arbitrariness of identities.

The Population

Foucault outlines how the modern state enhanced its power by intervening in the very life of the 'bio-politics of the population' (1980a, b, 139). Bio-politics leads to his overall perspective of politics or 'governmentality', 'the art of government'

(1991a, b, c, d, 90). In this process, power has two poles. First, a pole of transformation and second, the human body as an object of control and manipulation. The first revolves around the notion of 'scientific categorisation': for example, 'species' and 'population'. It is these categories that become object of systematic and sustained political intervention. The other pole is not 'human species' but the human body: not in its biological sense, but as an object of control and manipulation. Collectively, Foucault calls these procedures 'technologies' which centre around the 'objectification' of the body. The overall aim is to forge 'a docile body that may be subjected, used, transformed and improved' (1977a, b, c, 198).

Beginning with the inception of modernity, Western administrators rationalised their social problems with technically efficient means of population control: statistics, police, health regulations and centralised welfare. Such means constituted governmentality: an assemble of health and social care of ruling practices, knowledge authorities and moral imperatives that converged on the population in order to extend the reach of the state. The controversial point is that governmentality is more complex than state power. Custodial institutions and health programmes configured individuals into sub-strata of the population. For example, pension policies explicate 'the elderly' as a particular group of people, while statistics elaborate their status as a demographic entity (a 'health and social care population'). Thus, the disciplinary formation of subjects as a population makes possible the government of subjectification.

The Individual

If disciplinary gaze is a first step, then 'interiorisation' of that gaze is the second. Foucault's social constructivism, consisting of classification and dividing practices, technologies of the self and political grids of bodies and populations, has fuelled his critics claims that he deprives human subjectivity of health and social carency (Smart, 1983). Minson claims that Foucault burdens the body with being true subject of history and 'the flickering counterpart to the dull individual of sociology' (1985, 93).

Foucault emphasises two important aspects of individual that counteract his critics. First, the victims of modernity's disciplinary matrix – the prisoners, patients and children – can subvert the regulatory forms of knowledge and subjectivity imposed upon them. Second, while power/knowledge relations construct governable individual subjects, such subjects are not fixed to their conditions of ruling and do become health and social care of resistance to them (Foucault, 1977a, b, c, 1991a, b, c, d). To investigate the 'how' of power then requires: 'taking the forms of resistance against different forms of power as a starting point…it consists of using this resistance as a chemical catalyst so as to bring to light power relations, locate their position, find out their point of application and the methods used. Rather than analyzing power relations from the point of view of its internal rationality, it consists of analyzing power relations through the antagonism of strategies' (1980a, b, 211). Power is exercised on free subjects but does not necessarily determine conduct.

In this formulation, the individual is not the traditional subject caught in a war between domination and liberation. Rather, the individual is the personal space where both active and passive aspects of human health and social care and identity surface in the context of material practices. The production of identity is implicated in the production of power which is both positive and negative. Identity may be imposed through the surveillance of a subject population. This surveillance produces both discipline (i.e., conformity to the norm) and the disciplines (regulated fields of knowledge and expertise). Disciplinary surveillance involves first individualising each member of the population to facilitate the collation of observations across the population.

From these observations, statistical norms are produced relating to a multitude of characteristics. These norms are then applied back to the subjected individuals who are categorised, evaluated and acted upon according to their relation to the produced norm. Foucault's work focused on the 'history of the present' and 'power/knowledge' synthesis and how the subject was formed (Foucault, 1977a, b, c, 1978a, b, c, d). Here Foucault's work is on the 'microphysics of power' and the interplay of power relations, dividing practices and tactics in particular contexts (Foucault, 1977a, b, c): the 'doctor' and 'patient', 'prison officer' and 'prisoner', 'teacher' and 'student' and 'care manager and social carer' and 'career consumer'.

Implications

> It may be that the problem about the self does not have to do with discovering what it is, but maybe has to do with discovering that the self is nothing more than a correlate of technology built into our history. (Foucault, 1993, 222)

Foucault's formulation presumes the notion that individual lives are never quite complete and finished – that in order to function socially, individuals must somehow work on themselves to turn themselves into subjects. The notion of 'technologies' offers the opportunity for a particular analysis of the sites and methods whereby certain effects on the subject are brought about.

Objectifying technologies of control are, for example, those invented in conformity with the facets of self-understanding provided by criminality, sexuality, medicine and psychiatry investigated by Foucault. These are deployed within concrete institutional settings whose architecture testifies to the 'truth' of the objects they contain. Thus, the possibilities of self-experience on the part of the subject are in itself affected by the presence of someone who has the authority to decide that they are 'truly' ill such as a 'doctor' of medicine (Powell & Biggs, 2000). 'Subjectifying' technologies of self-control are those through which individuals: 'effect by their own means or with the help of others a certain number of operations on their own bodies and souls, thoughts, conduct and way of being, so as to transform themselves in order to attain a certain state of happiness, purity, wisdom, perfection or immortality' (Foucault, 1988, 18).

The important issues that Foucault raises via a questioning of the centrality of the subject are associated with 'truthful' formulations of the task or the problem that certain domains of experience and activity pose for individuals themselves. The boundaries of self-experience change with every acquisition, on the part of individuals, of a possibility, or a right, or an obligation, to state a certain 'truth' about themselves. For example, bio-technology in popular culture can tell a 'truth' of selling a dream of unspoken desire of 'not growing care' to people. However, it is the self-experience of subjects that can refute, deny and accept the 'truth' claims of bio-technology. In the case of lifestyles in popular culture, the active adoption of particular consumer practices, such as uses of bio-technology, contributes to a narrative that is compensatory in its construction of self (Biggs & Powell, 2001). Thus, the recourse to the notion of technologies of self is capable of accommodating the complexity of the 'subject'.

Although Foucault maintained the distinction between the technologies of power/domination and the technologies of self, these should not be regarded as acting in opposition to or in isolation to one another. Indeed, Foucault frequently spoke of the importance of considering the contingency of both in their interaction and interdependence, by identifying specific examples: 'the point where the technologies of domination of individuals over one another have recourse to processes by which the individual acts upon himself and, conversely, the points where the technologies of the self are integrated into structures of coercion' (Foucault, 1993, 203). The distinction should therefore be considered as a heuristic device and not the portrayal of two conflicting sets of interests. Overall, we should see Foucault's entire works as providing ways of understanding social relations that require on our part active interpretation, not passive regurgitation.

To take one modern example of how we might think with, alongside (and against perhaps?) Foucault, take the question: how is modern bioethics rooted in a specific configuration of subjectivity? The body culturally represents the best hiding place, a hiding place of internal illnesses that remains inconspicuous until the advent of 'expert' intervention. In other words, what are the effects of this problematisation given its conditions of possibility? Subjective relations to the self will be affected to the extent that social life confronts individuals with the proposition that this subjective truth – the truth of their relation to themselves and to others – may be revealed by 'bodies', which are also object of manipulation, transformation, desire and hope. In this way, we might anticipate through 'culture' (Morris, 1998) the relations between illnesses, new technologies, power, the body and desire. While confronting an illness, this involves a deliberate practice of self-transformation, and such tranformativity must pass through learning about the self from the truth by personal narratives within popular culture. How is this culture and the body itself, however, interacting with and being changed by advances in biomedical technology and the power of huge pharmaceutical companies?

Foucault is often seen as a structuralist, along with those such as Barthes, Althusser and Levi-Strauss. In reply to questions which sought to make such parallels, e was consistent: 'I am obliged to repeat it continually. I have never used any of the concept which can be considered characteristic of structuralism' (1989, 990). Perhaps the best way to view this is by examining his idea of historical 'events'. He refuses to see events as symptomatic of deeper social structures and focuses upon what seems to be marginal as indicative of relations of power. Events thereby differ in their capacity to produce effects. The following quote helps us see how this can be applied to cultural analysis: 'The problem is at once to distinguish among events, to differentiate the networks and levels to which they belong, and to reconstitute the lines along which they are connected and engender one another. From this follows a refusal of analyses couched in terms of the symbolic field or the domain of signifying structures, and a recourse to analyses in terms of the genealogy of relations of force, strategic development, and tactics. Here I believe one's point of reference should not be to the great model of language and signs, but to that of war and battle' (Foucault, 1980a, b, 114).

What about those questions concerned with whose culture, whose identity and how is this produced? These are the questions that pre-occupied Foucault. His refusal to see power as a property, of say a particular class, immediately leaves a question over his politics in terms of the idea of struggle. As he said: 'I label political everything that has to do with class struggle, and social everything that derives from and is a consequence of the class struggle, expressed in human relationships and in institutions' (1989, 104).

This leaves us with a question: against whom do we struggle if they are not the owners of power? Who creates cultures and how might alternative forms find public expression and does this change anything? These questions immediately bring forth issues concerning the relationship between Foucault and Marxist theory. Class structure, race and gender are key determinants of the position of individuals in capitalist society. It is difficult for 'techniques of resistance' to be mobilised when particular groups are de-commodified and marginalised and lose their social worth and voice (Biggs & Powell, 2001). At the same time, Foucault sees subjectivity not as a fabricated part of a deeper reality waiting to be uncovered, but an aspect of the reality systematically formulated by resistances and discourses. He sidesteps the binary relationship set up by Marxist theory between true and false realities and ways of knowing and political consciousness (Foucault, 1980a, b) and seeks to loosen knowledge, ideas and subject positions from categories of social totality: for example, social formation, mode of production, economy and society.

Culture is rearticulated in Foucault's thought to historical and societal features ignored in those models of social reality that 'read off' culture according to deeper structures. Foucault looks to areas such as medicine, sexuality, welfare, selfhood and the law, and to marginalised social groups, local politics and the micro-levels of culture. In these studies, he found social, discursive and historical substrata in which relations of domination were apparent that were not simply reducible to modes of economic exploitation. The idea of 'governing' then captures the ways in which the 'possible field of action of others' (Foucault, 1982, 221) are structured. Yet in inheriting this approach, authors have produced panoptic visions in which resistance is subsumed within impersonal forces. This results from over-looking two main aspects in Foucault's work. First, in terms of his own question, what are the 'limits of appropriation' of discourse'? Without this in place, all does appear quiet on the battleground. Second, and relatedly, the agonism that exists between power and freedom (May, 1999). This suggests that where there is power, there is also resistance; power thus presupposes a free subject. If there is no choice in actions, there is no power. A slave, therefore, is not in a power relationship, but one of physical constraint (Foucault, 1982).

Foucault notes three types of struggle: those against domination, those against exploitation and those against subjection and submission. The latter, while rising in importance in the contemporary era, do not do so to the exclusion of domination and exploitation as many of his followers have appeared to suggest. To understand why particular actors enjoy more power than others, as opposed to seeing power as a 'machine in which everyone is caught' (Foucault, 1980a, b, 156), an account of resistance is needed. Because Foucault views freedom as part of the exercise of power, he does not provide for such an account. Yet, in an answer to a question concerning 'power as evil', he spoke of the need to resist domination in everyday life: 'The problem is rather to know how you are to avoid these practices - where power cannot play and where it is not evil in itself' (Foucault, 1991b, 18).

What makes Foucault's overall theoretical work inspiring is how he animates and locates problems of knowledge as 'pieces' of the larger contest between modernity and its subjects. By downplaying the individual subject, Foucault shows how 'bodies' and 'populations' are sites were 'human beings are made subjects' by 'power/knowledge' practices (Smart, 1983, 44). To look for a possible form of transgression in order to change social relation, we must examine within contemporary arrangements the possibility for it to be 'otherwise'. We thus find, in Foucault's later work, an insistence upon the reversibility of discourses through 'resistance'. Subjects of power are also related to 'health and social care' who can strategically mobilise disjunctures in discourses and in so doing, open up the world of possibility in a world that seeks order through discipline and surveillance.

Chapter 3
A Foucauldian Analysis of Health and Social Care

Introduction

An interesting observation to make is that there has been little Foucauldian study of health and social care in relation to aging, and critical theorists have had to rely on a limited number of sources in the work of American and Canadian writers such as Katz (2007) and Frank (1998) and more recently from writers in the United Kingdom (Biggs & Powell, 1999; Powell & Biggs, 2000; Powell, 2005). Together with the United States, the United Kingdom has been subject to a succession of social policy trials, from welfare paternalism following the Second World War vis-à-vis William Beveridge through the Thatcher turn to marketisation to save public resources in health and social care and on into the Blair 'third way' endorsement of communitarianism. The focus on this is basically local communities become important to 'manage' old age was an interesting way to deviate the State away from care of older people through 'action at a distance' and focus on communities and families, professions and older people themselves for their own care. The current focus in the United Kingdom is on integrated health and social care but has led to a situation where health far outweighs social care in terms of priority, resources and staffing. While the aspiration is integrated, the reality is a disconnect in terms of governmental priority. A further concern is the Boris Johnson administration has talked up the importance of social care of older people but provided no consultation or even suggests they are unsure how it will be funded. This is especially contentious given that the UK government has committed billions of pounds on a new nuclear missile system (Phillipson, 2013). Such changes have had significant implications for the social discourses that impinge upon the social construction of aging and focus given to old age itself. The key question is how did we get here and where is the health and social care of older people traversing?

J. Powell, *New Perspectives on Health and Social Care*, International Perspectives on Social Policy, Administration, and Practice,
https://doi.org/10.1007/978-3-031-25432-1_3

The Biomedical Model and Aging

One of the startling continuities of dominant discourses on aging and old age has traditionally consisted of the construction of aging as a process of a disciplinary matrix of economic, social and physical decline. In occidental culture in particular, the aging body is perceived to be the 'bottom line', subjecting us to relentless 'betrayals' of the human body through physical deterioration and the consequent tasks of maintenance (Dittman-Kolhi, 1991) and compensation (Baltes & Carlestensen, 1996). Insofar as there is a history of aging, there is also a history of efforts to control, supervise and self-regulate the aging body.

The consequences of the decline discourse can be seen most clearly in the dominance of medico-technical solutions to the problems that aging is thought to pose. Older people as subjects rather than as people play into the ageist stereotype of how older people are not just viewed but treated in terms of care. This, according to Katz (2007), has led to a fundamental bias of gerontological theorising and research towards geriatric medicine and the relative failure of more broadly based life-course approaches to impinge upon engaging and therefore old age. There have been some important exceptions to this trend, most notably in attempts to develop a humanistic gerontology (Cole et al., 1992), and a flirtation with the impacts of postmodernity on aging identity, based primarily on the United States and United Kingdom. However, it is arguable whether these debates have had a broader impact on everyday perceptions of age and aging. Indeed, a preoccupation with the medical challenges presented by aging, underpinned by privatised and insurance-driven health provisions, has resulted in what Michel Foucault (1973) may have observed as an expansion of the medical 'gaze' into all areas of social policy. The medical 'gaze' refers here to discourses, languages and ways of seeing that shape the understanding of aging into questions that centre on, and increase the power of, the health professions, in particular, and restrict or delegitimise other imagined, strategic or real possibilities. A consequence is that areas of policy that may at first seem tangential to the medical project come to be reflected in its particular distorting mirror. The impingement of the medical gaze can be seen in recent policy debates about the consequences of Brexit concerning disadvantaged groups over a shrinking public purse, decline of GDP and fears of a breakdown of an intergenerational social contract, considered to be a foundation of post-war welfare policy (Phillipson, 1998). The impact of medicalised notions of aging and its construction as a threat to other sections of the population can also be seen in Moody's (1998) critique of bioethics and aging and the proposed rationing of Medicare coverage in American welfare policy. Here, medical care has come both to colonise notions of old age and to reinforce ageist social prejudices to the extent that infirmity has come to stand for the process of aging itself and medicine its potential facilitator.

A key point here is that the notion of a 'medical gaze', as first described by Foucault, not only draws our attention to the ways that aging has become 'medicalised' as a social issue but also highlights the way in which older people are encouraged for as long as possible to 'work on themselves' as active subjects. Thus, as Blaikie (1999) has pointed out, older citizens are encouraged to take greater personal responsibility for their health and for extending this period of their aging. Those who move into a fourth age, defined as a relative failure of that 'work' and direct dependence on health and welfare services, then discover themselves transformed into passive subjects who have a voice but not (and in some cases, never) listened to (Powell & Biggs, 2000).

While significant sections of public debate on aging have been rooted in this biomedical gaze, the growth of an able-bodied and pension-rich cohort of older people has led to a parallel discourse on consumer-aging – much blame of the 52% Brexit vote in the United Kingdom has been blamed on older people, while no evidence has been provided. The consequences for theorists such as Katz (1998) have catalogued the rise of the older consumer who are rich and an identification of a 'grey market' for consumer goods and services. This gives a misleading view that all older people are rich and are at fault for public policy decisions and votes associated with the future of UK's relationship with the EU. Discourses of consumption and having huge resources can pull from theoretical traditions such as activity theory (Blaikie, 1999) and more recent postmodern interpretations of aging lifestyles which fail to take into account poverty in old age (Phillipson, 2013) which causes ill health requiring health and social care services. In poverty, to have a brutal choice of eat or heat based on minimum resources has not been researched sufficiently.

It appears, at least in the United Kingdom, that established and emerging 'master narratives' of biological decline on the one hand and alleged being a consumer who is rich on the other coexist, talking to different populations which underplays the real consequences of low pensions and a care system that is not fixed. This is contradictory as one can see.

They are contradictory in their relation to notions of autonomy, independence and dependency on others, yet linked through the importance of the adoption of 'golden-age' lifestyles of rich 'baby boomers'. However, this focus on medicalisation and lifestyle has tended to obscure a third discourse on aging that has traditionally been strong in Europe, the United Kingdom and Australia, namely, the association between old age and health and social care.

This association needs unpacking with a strong theoretical analytical base to draw out the implications of health and social care for older people. Drawing from Michel Foucault, his conceptual excavations and methodologies will be utilised to discourses of care and their related technologies embedded in social work that have emerged in relation to successive stages of aging as an issue in social care policy. This focus on care as a factor in the reconstruction of contemporary aging will draw most heavily on the areas with which we are most familiar, namely, UK's manifestations of wider global trends in care policy.

Foucault and Aging: An Important Conceptual Toolkit

Michel Foucault's theoretical challenges are always posed in complicated, complex and multifaceted terms. He urges people to 'refuse what we are' (1982, 216), meaning that we should refuse to remain tied to fixed identities to which older people, for example, are subjected – having the ability to change one's own identity. He linked his own project with all those who struggle against the ways in which they are individualised, particularised and objectified by controlling discourses.

The main point of Foucault's methodology, called a genealogy because of its emphasis on tracing historical pathways that have contributed to contemporary circumstances, was to identify discourses. His concept of 'discourse' is a key term both in understanding Foucault's work and in explaining facets of care. Foucault identified discourses as historically variable ways of specifying knowledge and truth. They function as sets of rules, and the exercise of these rules and discourses in programmes that specify what is or is not the case – what constitutes 'old age', for example. Those who are labelled 'old' are in the grip of power. This power would include that operated by professionals through institutions and face-to-face interactions with their patients and clients. Power is constituted in discourses, and it is in discourses such as those of 'social work' that power lies. Genealogy is concerned, then, with the historical limits and conditions of socially determined discourses, which then direct and distort the personal and institutional narratives that can subsist within them. When a discourse has stabilised historically, it can be referred to as a 'discursive formation', which can come to characterise a particular period of welfare development and the associated possibilities for identity performance that it contains (Foucault, 1977a, b, c).

Michel Foucault (1967) was particularly interested in the limits and potentials of discourses from 'human sciences' because of their attempts to define human subjectivity. His attention shifts to the power of professionals because Foucault found that the conditions of potential for 'true' discourses about human subjects include complex relations between knowledge about people and systems of power. Here Foucault focuses on the techniques of power/knowledge that operate within an institution and that simultaneously create 'a whole domain of knowledge and a whole type of power' (1977a, b, c, 185). These domains effectively destroy the legitimacy of other, competing, discourses, just as a professional medical opinion might de-legitimise voices arising from folk medicine or informal care. The genealogical work of uncloaking these power relations is characterised, by Foucault, as setting out the 'political regime of the production of truth' (Davidson, 1986, 224).

The effects of the reflexive relationship between power and knowledge that is implied here would include the tendency for 'professional power' (espoused by health and social care professions) to be reinforced by the sorts of questions professionals ask and the data they collect – in the main known as 'evidence-based research' which undermines qualitative research from the narratives of patients or older people. This hegemonic knowledge then progresses to a certain definition of a problem area that then feeds back to stabilise the original formulation of the 'problem' itself. By the same token, diverse public policy positions point such professionals to seek out certain forms of knowledge that tend to reinforce the assumption base of the position of that public policy and its associated discourses. As part of this process, certain powerful voices increase their legitimacy, while other, often dissenting, voices become irrelevant.

A consequence of the mutually reinforcing relationship between power and knowledge that emerges from the above is to construct older people concurrently as subjects and as objects. First, people are seen as objects by someone else, through control and restraint. Second, people are deemed to actively subject their own identity to personal direction through processes such as conscience and mediated self-knowledge. Foucault (1988) refers to this second process as 'technologies of self'. Foucault's formulation of 'technologies of self' claims that individual lives are never quite complete and finished – that in order to function socially, individuals must somehow work on themselves to turn themselves into subjects. The notion of 'technologies' offers the scope for an analysis of the sites whereby certain effects on old age are brought about. As Foucault puts it: 'Both meanings [of control and self-conscience] suggest a form of power which subjugates and makes subject to' (1982, 212).

In terms of care, itself a discourse, both clients and social workers would need not simply to follow the rules that legitimise what they can say and do, but also to work on themselves so each can become the sort of person who can be seen and heard within that discourse. If they are not careful, both professionals and users of health and social care systems become trapped in a dance of mutually maintained positions that serves to sustain a particular view of aging and the remedies, the technologies, that can be brought to bear on it (Phillipson, 2013). The question then is how do we 'dig' underneath such powerful discourses that open up the relationship based on trust and reciprocity rather than agism and assumption-based relationships?

An analysis of the contested notion of power itself, which follows the methodological pathway as mapped out by Michel Foucault, must assess three aspects of how such power is created and maintained. First, the analysis must examine the genealogy of existing relations, how they have emerged and the discourses they both reflect and reinforce with respect to aging as seen as both objects and subjects. Second, attention must be given to the broad distribution of power and knowledge that these relations imply – this is difficult to capture but a challenge a health researcher must investigate and disseminate. Lastly, technologies of care such as care management will need to be critically assessed as approaches to the self that hold certain webs of power in place. Each will contribute to the ways in which older people as subjects enmeshed in certain relations apply techniques of identity control to themselves. This needs historically unpacking and what the policy and practice implications are for the care and positioning of old age.

A History of the Present: The Management of Aging by Social Work

The past 200 years has witnessed an increasing institutionalisation of the life stage of 'old age', both in the United States (Chudacoff, 1989) and in the United Kingdom. The emergence of what could be called 'professional power' in what has since come to be called modernity is also associated with transformations that took place from the nineteenth century onwards. In the case of care, these transformations have been associated with a series of moral panics about the family in which the state was expected to intervene (Jones, 1983). Professional social work developed as a hybrid in this space between the public and private spheres and was produced by new sets of relations among the law, administration, medicine, the school and the family.

The rise and alliance of social work was seen as a 'benevolent' solution to a major problem posed to the State; namely, how can the State establish the health and development of family members who are 'dependent' while promoting the family as the 'natural' sphere for caring for those individuals and thus not intervening in all families? Thus, social work developed between individual families, older people and the State, which would be in risk of taking responsibility for everybody's human needs and hence undermining the responsibility and role of the family. Did the family need such intervention? This somewhat ambivalent positioning of a new discipline of 'social work' meant that from its beginning, the social work profession has had to carefully negotiate the boundary between public expectation and private conduct – an enormously difficult task. As such, the social work project has disguised with a double perspective of external coercion of family life and 'personal cure' (Phillipson, 2013) as it embraced both the judicial and the therapeutic in acts of intervention in older people's lives.

There can be little doubt that much of the traditional identity of professional social work rests on what can be identified as 'modern' foundations of the formation of society. While nursing and medicine have drawn heavily on technical/scientific knowledge to justify their legitimate status, social work has drawn, with relative degrees of success and in succeeding periods, on arguably both psychoanalysis and applied social sciences. Both health and social care have been part of a great movement for 'progress' characteristic of the twentieth-century 'grand narrative'. Of intervening in family lives. The key question, as Phillipson (1998) acutely observes, is this the case of all families or families who have been problematised as needing intervention because members may have been unemployed or in poverty. Would the same tenacity of intervention have been the same for rich families and their older siblings?

As the twentieth century has proceeded, the growth of professional social work has become increasingly dependent upon its interrelationships with the welfare state, which provided its primary rationale and legitimacy to intervene and cement its power base. As a consequence, social work mediated not only between potentially socially excluded individuals and the State but also with diverse private and voluntary agencies. Further, social work became closely related to the development of new forms of social regulation associated with the increased complexity of modern society. To put it another way, these new forms of social surveillance were characterised by notions of normalisation, discipline and surveillance (Foucault, 1977a, b, c). For Foucault, such arrangements come to constitute a 'total set of relations that unite at a given period, the discursive practices that give rise to epistemological figures, sciences and possibly formalized systems' (Foucault, 1972, 191). They systematise networks of ideas about the 'nature' of individuals, their perfectibility, the reasons for their behaviour and the ways they may be classified, selected and controlled. Social work became an instrument with which to manage individuals by the manipulation of their qualities and attributes, depending on applied social-scientific knowledge and professional expertise.

Under particular social circumstances and as history changed, discourses emerged that both lead to the creation of new professions that in turn simultaneously reinforced the discourse itself. In so doing, the development of a new type of knowledge about older people emerged, and new sites in which to grow old were created.

A Genealogy of the Post-Second World War Consensus of Care and Aging

In the era following the Second World War, which saw the consolidation of care systems in much of the occidental world, old age also came to be seen as problematic in a certain way. Williams Beveridge (1942, cited in Wilson, 1991), for example, whom some credit as being the architect of the British welfare state, says of older people:

> It is dangerous to be in any way lavish to old age until adequate provision has been assured for all other vital needs, such as the prevention of disease and the adequate nutrition of the young. (Beveridge, quoted in Wilson, 1991, 39)

The quote is clear. Children were to be the priority despite the universalism that the welfare state would be there from the cradle to the grave irrespective of age – which was demolished as mythical. Worse, while on the one hand, the older person was portrayed as a stoical and heroic survivor in the immediate post-war period, this representation was contingent on an absence of demand on the rest of society. This ambivalence was reinforced by the difficulty of reconciling old age with the rhetoric of progress and investment for the future that characterised the growth and ideological justification of children. Nor did older people fit narratives of care but add production, work and usefulness to capitalist production, used to justify welfare in terms of maintaining the current workforce (Phillipson, 2013).

When older people came to the attention of social work in this period, intervention was allowed when the conduct of an older person was judged to be a danger to him or herself or to others, most notably as a 'health hazard'. In fact, associating social work with the future, social investment and protection from social hazards had contributed to a deep embedding of the coupling of later life as a stereotypical burden on society.

Here, this social work discourse of aging and its positioning as a medicalised yet mythical story of decline and maintenance that dominates much of the Western literature on aging. In the post-war consensus on health and social care, old age took on a double and somewhat contradictory character: the pension-worthy survivor of the war and the burdensome hazard to society.

The new welfare state, and its associated 'welfare gaze', simply did not see them as people. In both senses, social welfare came to colonise the meaning given to old age in the public imagination, and the welfare state and its care professions came to characterise the place, the discourse, in relationship to which aging identities have come to be formed: decline and stigma of lost personal control.

The next section illuminates the technologies embedded in social work practice that exemplify this phase of care: psycho-casework with older people.

Psycho-Casework with Older People

Since the 1960s, the new human sciences had as their central aim the prediction of future behaviour (Ignatieff, 1978), which fit well with social work's professional mission and what emerged as its chosen method: psychoanalysis (Lubove, 1966). Psychoanalysis supplied a language and way of thinking, which served to pathologise older age and also happened to suit the needs of social production. The negative stance taken by psychoanalysis to older age has been catalogued elsewhere (Biggs, 1999). The point here is that this negative stance, coupled with the need for a discourse that both reinforced professional power and the marginal positioning of older people, found each other in the early use of psychodynamic language by social workers on both sides of the Atlantic. Thus, the 'caring' profession tended to draw upon psychoanalytic discourse to socially construct an image of older clients as 'greedy and demanding, always clamoring for material help, always complaining of unfair treatment or deprivation; this attitude shades into paranoid imagining' (Irvine, 1954, 27). This psychologised view of failing independence closely parallels an economic discourse that old age constitutes a drain on resources that could be used more 'productively'. As another powerful discourse, the conceptualisation of age as burden has developed an enduring presence. In a survey of social service departments 20 years later, Satyamurti (1974, 9) observed:

> The language that social workers use about their clients, often jokingly, seems often still to be based on an image of them as good or naughty . . . It seemed, too, that when social workers referred to a 'difficult case' they did not mean that the client presented problems that were difficult to solve, but that he was demanding and time-consuming.

Discourses of 'dependency' formed the foundations of practice development in modern society in relation to older people. The notion of dependency was articulated in terms of policy through the state provision of care services and via social work through the practice of care. Rather than valuing older people, they were devalued. Rather than empowering older people, they were disempowered.

This positioning was reinforced as knowledge was collected on older populations throughout such agencies and remedies channelled through their offices. Psychoanalytic thinking, rather than an occasion for individual liberation, became the language and the technique through which the identities of professionals and their older clients were shaped. This seeped into society and created further ambivalence to older people.

Health, Care, Market Forces and Aging

With the marketisation of welfare, this psychological point has itself been significantly eroded, and with it the traditional role of social worker as provider and counsellor. Not only do new discourses provide a 'swarming' of professional power/knowledge, but they can also take away. And in the Reagan and Thatcher years of the late 1980s and early 1990s, social work had to reinvent itself – and its construal of older people – in order to survive.

Controversially, old age became, in this period, increasingly associated with risk, both personal and structural and, at the same time, was subject to a privatisation of that risk and a withdrawal or rolling back of supports, previously taken as stable and enduring. It is not by chance that an increased focus on risk in social work has coincided with the decline in trust in social workers' expertise, decision-making through psychoanalytical insights and a growing reliance on increasingly complex systems of managerialism with older people themselves as 'consumers' (not all) of services. Such a growth has constituted a conducive framework based on the language of the market and its pragmatic management as opposed to trust in applied social-scientific discourse, although preceding emphases on the psychoanalytic can be brought in from time to time to reinforce an individualised notion of personal responsibility. New policy priorities require new technologies if they are to influence the control of resources of their subjects and objects. The new technique introduced to UK and Australasian welfare was care management (Powell, 2005). However, rather than being an attempt to co-ordinate an already privatised and fragmented welfare system as existed in the United States, care management United Kingdom-style was used as a mechanism to deconstruct the existing state-run system and introduce a marketised care economy. It privatised care.

Aging and Care Management

The role of the care manager in the United Kingdom over the past 30 years to the present marked a fundamental shift in social work from a practice-based to a managerial role and identity. As a result, the management and delivery of care has become increasingly indirect. It has become indirect in two ways: first, the pivotal function of the case manager is seen as the co-ordination of packages of care that draw on services provided by private and not-for-profit agencies; second, there is an increased emphasis on assessment and the monitoring of standards of those services supplied by others. Indeed, in this regard, one can cite Howe's (1994) view that the managerial role highlights a shift from social welfare to surveillance and control. In emergent managerialised regimes, judgement is increasingly bound up with managerial necessities concerning corporate objectives and resource 'control' (Flynn, 1992). The devolution of managerial responsibilities is intended to turn clients into consumers and to constrain professional autonomy by having such managers internalise budgetary disciplines. A result is that managerial processes and categories of assessment are compounded with other categories of 'risk' (Beck, 1992) and in some cases supplant them almost entirely (Phillipson, 2013).

Care management makes sense as part of a discourse that displaces and reduces the financial 'burden' of age on the state and onto the families of vulnerable older people. Economic privatisation is accompanied by a wish to see those same older people as active consumers, making choices between services and changing services or residence if they are found wanting – from their own resources – some without hardly any. Hence, there has been little consideration, however, of the financial costs, the costs to well-being or the ability of such vulnerable groups to act in accordance within a discourse based on consumption.

While care management has proved an effective technology for transforming welfare economies, it has made little sense in terms of the preceding social work ethos of counselling and direct care. It is here that the second 'surveillance' aspect of case management technology has come to the fore as a source of professional power/knowledge. Social workers are now the risk-assessors and enforcers of a mixed care economy, a discourse that leaves older people who use services on the contradictory and risky ground of being simultaneously consumers and potential victims being in poverty. It is this contradiction that the Thatcher/Reagan turn in political discourse had given its social democratic governmental successors.

The Embedded Market of Care and Aging

An unwillingness to increase public finance for older people has led both the Blair and Clinton administrations to leave the market-welfare systems of both countries relatively untouched. However, it is possible, at least in the United Kingdom, to observe a change in the rhetoric, and by association the policy discourse, legitimising the place of older users within welfare services.

Using communities to engage older people changed the discourse of care management yet again. It has become a priority on the one hand to 'include' older people back into the wider social fabric as 'active' participants, and on the other hand, to protect those who are sufficiently infirm not to be able to participate. This contradictory inheritance had led to two parallel and independent policy initiatives.

First, government-sponsored initiatives such as 'Better Government for Older People' (1998) have been used to promote short-term partnerships with service providers. Second, a debate on the nature of mental incapacity (in other words, when older citizens are judged no longer capable of existing under the rubric of partnership) has resulted in policy guidance entitled 'No Secrets: The Protection of Vulnerable Adults', which draws on an increasingly inquisitive version of care management a version of professional social work not old age as a legitimate voice (Powell, 2005).

Conclusion

The recognition of partnership in communities appears could mark a shift away from the traditional role of policy as facilitating progressive disengagement and dependency. It also links policy with changes in the lifestyles and self-governance adopted by older people themselves, listening to them and acting on their experiences. This is essential as much of the focus is on the legitimacy and reconstruction of professional services in health and social care rather than listening to the real experts – older people themselves. This impacts on risk.

However, these developments have their dark side, and the ethics of using such social work technologies to deny the force of aging as a human experience have been subject to less scrutiny as espoused by the chapter. Indeed, it is perhaps emblematic of contemporary occidental culture that each of the shifting social work identities identified above and different ways to 'manage' the care of older people offers the promise of escape from, rather than a deepened understanding of, aging experiences. Those who do not conform to the social work framework appear to have been shunted into a non-participative discourse, bounded by professional surveillance, or the more edible yet closely related discourse of 'monitoring'. In both cases, it could be suggested that a discourse on dependency driven by the post-Second World War has been accompanied, and in some cases replaced, by a discourse on risk. This stretches to the present and post Brexit. The risk of giving into an aging body, the risk of thereby being excluded from one's retirement community, the risk of being too poor to maintain a consumer lifestyle, the risk of being excluded from participation through incapacity that has been externally assessed by social workers, the risk of being abused, the risk of control being taken out of one's hands, the risk of tokenism in partnership and worse of all, the risk of being ignored – especially when care is crucial for older people as individuals and as populations.

A final yet crucial point, and one that links a Foucauldian assessment with contemporary trends in understanding modern aging in health and social care, is to suggest that a Foucauldian analyses of discourse and power explicate what it means to speak of narratives of aging. It suggests that narratives are not personal fictions by which older people choose to live by, but are discourses subject to social, economic and historical influence by external forces and powerful biomedical assumptions. Narratives of aging are personal in so far as older people apply techniques to themselves, while the professional technologies and the knowledge base on which they are legitimised imply particular power differentials that will determine the way and the what of the storyline of how society treats older people who may require health and social care services and how they are managed.

Chapter 4
Surveillance and Health and Social Care

Introduction

The previous chapter explored a genealogy of health and social care transitions in the United Kingdom. The thesis pursued in this particular chapter is that an accelerating interest in personalisation is central to understanding modern care policy as a social phenomenon in occidental societies. It will be argued that personalisation legitimates practice in which the state monitors and co-ordinates but does not intervene. This has led to a social situation that has radically transformed social welfare of its traditional rationale as 'caregiver'. Simultaneously, informal care has become the centrepiece of social policy following the adoption of personalisation policies in the United Kingdom. One intended consequence of these policies has been to transfer the financial and emotional responsibilities for care to service users and informal carers under the aegis of 'personalisation'.

His sudden concern for the safety and financial security of career people, who are service users, legitimates a role for welfare professionals within the landscape constructed by personalisation policy. The price to be paid, however, is that the relationship between the State and career people has been reduced to one of surveillance and the enforcement of an oppressive notion of what community obligation might entail. As with other forms of tacit control, the surveillance role left to a residual local state evokes a 'surface' of reality as constructed as 'depth', whereby generic methods of surveillance are presented as 'concern' models. This act of observation confers a uniformity that emphasises the 'protective' role of the professional rather than the substantive requirements of career people at the centre of inspection.

The specific focus of this paper will be to make out how the generic concerns of personalisation policy have legitimised generic surveillance practices. Simplistically, this idea can be presented as a mathematical metaphor of 'inspection minus intervention equals surveillance'. It also serves to illuminate otherwise hidden facets

J. Powell, *New Perspectives on Health and Social Care*, International Perspectives on Social Policy, Administration, and Practice,
https://doi.org/10.1007/978-3-031-25432-1_4

affecting the practice of personalisation, which while based on foundational concerns can unwittingly serve more adverse purposes.

The Rise of Personalisation

In the United Kingdom, formal recognition of personalisation has occurred with two policy initiatives: first, the implementation of the health policy under the then British Conservative administration under John Major (1992–1997). This included moves to enforce a privatised welfare economy, with emphasis on the centrality of informal care (Powell, 2006a, b). A residual role for the local state included monitoring and inspecting care purchased from the 'mixed economy'. Indeed, the development under Tony Blair and 'New Labour' in the late 1990s of a 'third way' campaign that sought to reinforce startling continuities of the previous administration of a moral obligation to care by communities or families. Similarly, David Cameron, the current British Prime Minister (2010–), has focused on the idea of a "Big Society" that has much in common with former UK Tony Blair's (1997–2007) government: the common aim of these policy narratives would be to transfer financial and caring responsibility for dependent career people away from the collective and to individual families. In other words, the recent issue of personalisation has reached professional salience at a time when the relationship between formal and informal care was being re-structured. The role of professional workers and the nature of informal care were both contested and in a state of considerable flux and uncertainty.

It will be argued that these two social policy initiatives have a number of common threads which establish a shift in career people's services away from care and support and towards the surveillance of those being cared for. The form that this shift has taken varies depending upon the site of interaction and subsequent power relations between professional workers and service users. For mental health services, surveillance is directly aimed at the nominated 'consumer' or 'patient'. It has led to the compulsory treatment of people in community settings and the instigation of community supervision orders.

Increased surveillance is often presented in social policy as a tactical response to crises at margins of personalisation policy, the accidental accretion of responses to unintended consequences. The argument pursued here will suggest that increased surveillance is part of a strategic process to engender control and morality.

It is not that personalisation has made more of an awareness of the fragmented variants of social care, but that personalisation gives meaning to care, and before its advent, technologies such as care health and social care were the welfare equivalent of a solution looking for a problem. Personalisation, in particular, fills a vacuum at the centre of social care policy, giving it an ideological legitimisation function it had previously not had, a policy flag for Cameron to hide behind in terms of ideology and cuts in public services.

Foucault and Surveillance

This chapter will explore issues in a number of ways. First, the methodological 'box of tools' drawn from the work of Michel Foucault (1977a, b, c) will be used to expand upon discontinuities between social policy and its consequences. Two themes will then be expanded, firstly, questions of morality to highlight change and the social policy technology available to execute it, namely, care health and social care. Secondly, the relationship between overt concerns and covert consequences will be analysed in order to examine how benevolent intentions, without critical analysis, can result in negative outcomes for the recipients of state intervention.

Foucault's main interest is in the ways in which individuals are constructed as social subjects, knowable through disciplines and discourses. The aim of Foucault's work has been to 'create a history of the different modes by which, in our culture human beings are made subjects' (1982: 208). In *Madness and Civilization* (1965), Foucault traces changes in the ways in which physical and mental illness was spoken about. Foucault employs a distinctive methodology for these studies, archaeology, which aims to provide a 'history of statements that claim the status of truth' (Davidson, 1986: 221). Foucault's later work, *Discipline and Punish*, focuses on the techniques of power that operate within an institution and which simultaneously create 'a whole domain of knowledge and type of power' (Foucault, 1977a, b, c: 185). This work is characterised as genealogy and sets out to examine the 'political regime of the production of truth' (Davidson, 1986: 224). Both archaeology and genealogy are concerned with the limits and conditions of discourses, but the latter takes into account political and economic concerns relevant to personalisation policy.

Indeed, the work of Foucault has engendered an awareness that modern institutions operate according to logics that are often at excessive variance with the humanist visions embedded in policy analysis (Penna & O'Brien, 1998). In other words, the overt meanings given to a certain policy of activity may not correspond to their consequences. Whether these outcomes are intended or accidental was less important to Foucault than the analysis of power. As Smart (1985: 77) points out, Foucauldian analysis asks of power: 'how is it exercised; by what means?' and second, 'what are the effects of the exercise of power?' Within those strategies, investigation would need to be centred on the mechanisms and the 'technologies' employed and to the consequences of any social momentum for change.

An example of the discordance between social policy, the philosophy that overtly drove a certain initiative, and its effects comes from Foucault's (1977a, b, c: 201) analysis of utilitarianism. Indeed, a pervasive theme of Foucault's (1977a, b, c) work is the way in which the panopticon technique 'would make it possible for a single gaze to see everything perfectly' (1977a, b, c: 173). Foucault describes how panopticism (based on the design of Jeremy Bentham) becomes a process whereby certain mechanisms permeate social systems beyond actual, physical institutions. Techniques are thus 'broken down into flexible methods of control, which may be transferred and adapted ... (as)... centres of observation disseminated throughout society' (1977a, b, c: 211–2).

The mechanisms used to extend the reach of centres of power will vary depending upon the ground upon which they are required to operate. Their function is to evoke and sustain moral interpretations of particular social behaviours throughout intermittent observation such that their objects come to internalise their own surveillance.

One important facet of Foucauldian analysis is the author's preoccupation with historical periods in which conventional values are in flux as in the case of madness, discipline and sexuality (Foucault, 1965, 1977a, b, c, 1978a, b, c, d) and how the emergence of professional discourses interpenetrate the evolution of new commonsensical understandings of 'normality'. There are, in other words, periods in which particular sites of control, for example, institutional care, family relations and intimate relationships, are subject to novel mechanisms and technologies in order to facilitate the transition from one state of affairs to another. These technologies may be overtly applied during periods of flux until moral relations have been accepted, and, during the process of their application, they both modify and are modified by the professional groupings charged with their implementation. While Foucault does not impose any sense of causality on the development of such discourses, it is possible to discern the need for both an explicit moral reason and a method of operation, shaped to whatever new contexts are appropriate. Government morality would act as a permissor for activities such as surveillance. A professional technology would provide a means of implementation depending upon the site (i.e. in institutions of the state) of the targeted activity.

As Rouse (1994) has pointed out, an examination of the relationship between power and knowledge is central to interpret and understand social phenomena through a Foucauldian gaze. This is particularly apposite where there is an attempt of a disaggregation of a stated policy and its mechanisms in order to discover what is thereby hidden or obscured. One of the consequences of power and knowledge is that rather than the focus on the explicit use of a particular technique of knowledge by someone in power to cause a certain effect, attention is drawn to the reflexive relationship between both elements. There is a concern then:

> with the epistemic context within which those bodies of knowledge becomes intelligible and authoritative. How statements were organised thematically, which of those statements counted as serious, who was empowered to speak seriously, and what questions and procedures were relevant to assess the credibility of those statements that were taken seriously. ..The types of objects in their domains were not already demarcated, but came into existence only contemporaneous with the discursive formations that made it possible to talk about them. (Rouse, 1994: 93)

So, just as knowledge shapes what action is possible, what power is exercised, those actions shape the creation of new knowledge and what is thereby given credence. Over time, legitimate 'domains' are established which both define what is real and what can be done about it. Other possible interpretations are simultaneously discounted and delegitimised. The result is a self-contained common-sense world in which power and knowledge support each other. These domains, for example, not only sustain certain professional discourses, but they also mould what those professions might become. This analysis of power and knowledge emphasises their entwinement and the processes that occur as a particular domain takes shape. It also marks a distinction between what a method for obtaining knowledge produces and the relationship between the shaping of that product and the distribution of power.

Returning to an earlier theme, the process by which a particular domain is established may not be the same as the reasoning given to explain what events take place and their effects. Indeed, as his understanding of this relationship developed, Foucault (1982: 86) indicated that 'power is tolerable only on condition that it mask a substantial part of itself. Its success is proportional to its ability to hide its own mechanisms'.

Furthermore, in personalisation policy, there is an open intention to 'empower' through allowing career people to live in their own communities and monitoring support, which may have become a means of policing informal care and through that the conduct of career people. Throughout the past 15 years, community care policy has drawn upon a number of sources of flux to achieve momentum. These have included a concern over familial obligation to care and changing social work practices, from a traditional providing role to that of health and social care and purchasing of services, to movements from directed care to personal budgets. However, the 'no-cost' option of a social policy reliant on personal budgets comes to look increasingly fragile. It is therefore in need of a shift to the moral ground of obligation and personalisation into which the Cameron administration continually attempts to tap through the 'Big Society'.

However, another complicating factor to an 'obligation'-based personalisation and social policy manifests from the idea that informal care is at root a voluntary activity. It is not, therefore, bound by any formal code of social practice, as would be the case for paid workers. Hence, there is no formal reason for intervention should a policy of informal care meet resistance. Thus, a social policy exists that contains fiscal policy and morality and makes informal care a legitimate responsibility. The threat of 'personalisation' provides the excuse for this invasion of the private sphere, a shift from 'consent' to 'coercion' and from 'support' to 'surveillance'. However, to be fully effective, a technology would also need to be found that would implement the logic of that policy.

The Technologies of Health and Social Care

The core technology by which community care can be implemented exists in the role of care management. It can be conceptualised as the co-ordination of services into a 'package and social care of care' in order to maintain 'clients' in community settings. The technology is indirect in three ways. First is the pivotal function of social care that draws on services made available through a 'mixed economy of welfare'. Second, there is a shift towards supporting informal carers rather than directly working with the nominated client. Third, there is the emphasis on assessment and monitoring of provision that is supplied by service providers.

This quality of indirectness 'makes sense' as a means of health and social care a 'mixed economy of welfare' which requires that those who purchase care, or their health and social care, are separated from those who provide it. Because of the intensification of marketisation, this limits the development of cartels and allows purchasers to choose between competing alternatives, thus placing them in the role of 'honest brokers' who assess need, supply information on the alternatives and then co-ordinate purchases. It does not, however, make sense in terms of direct care, intervention or interaction between career people and social workers other than as a sort of 'professional travel health and social care', advising clientele on the options, best deals and cash options. Care assessment and monitoring have now become an integral feature of social work practice and reflect a trend towards justifying welfare activities in terms of quality assurance.

The Department of Health has commissioned research into demonstration projects, in order to monitor the implementation of care health and social care. Unsurprisingly, this programme has generated scarce critical analysis of the role of this technology of care, focusing, instead, on the financial success achieved in its adoption. By replacing direct intervention with health and social care systems, the technology fails to provide guiding theoretical principles for interpreting and acting on conflict in social relationships. Smith and Brown (1992: 685) compound this by arguing 'one of the reasons for confusion around community care is the lack of an ideology to guide and influence its implementation'. They suggest that normalisation may provide such a role once a perspective on the duality of power and powerlessness has been developed. 'Techniques of resistance' (Foucault, 1977a, b, c) by career people to health and social care techniques were found by Powell (2006a, b: 12) who claimed career people 'were particularly adamant that they did not want to be "cases" and no-one needed to "health and social care" their lives'.

However, despite this resistance, the introduction of the 'mixed economy of welfare' in the United Kingdom has consequences for the surveillance of career people. The mixed economy reflects political rationalities and technologies of government. Welfare pluralism is used to mobilise the use of resources – and thereby embody power relations – and thereby supply an economic vocabulary to legitimise the allocation of those resources and associated schemes of inspection and surveillance of services for career people. Powell and Chamberlain (2012: 26) note that 'social actors', such as care health and social carers, try to translate values into their own terms, to provide standards for their own actions and in so doing, facilitate 'rule at a distance'. A mixed economy of welfare is a means of doing this; it fabricates representations of 'empowerment' for career people. As Chau (1995: 111–145) points out, not dissimilar to the social construction of healthcare accounting software, services become devices which transform real relations. In a sense, 'career people' become 'consumers', 'social workers' become 'health and social carers' and 'social service departments' become 'purchasers', all crystallised by the formation of community care policies. In this case, services provide schemas for the 'conduct of conduct' (Foucault, 1991a, b, c, d) dominated by power/knowledge and characterised by the discretionary autonomy of care health and social carers. It is within this disciplinary matrix of policy, practice and autonomy that power operates on career people, ultimately reinforcing the fragmentation that surveillance engenders in the psyches of career people at the centre of the professionals' gaze. This form of surveillance:

> clearly indicates the appearance of a new modality of power in which each individual receives as his status his own individuality, and in which he is linked by his status to the features, the measurements and gaps, the 'marks' that characterise...and make him a 'case' (Foucault, 1977a, b, c: 192).

Hence, the career client is marked out for perpetual surveillance throughout the remainder of his or her community care service. Carers and professionals also come under scrutiny as part of the continuous review of the client's needs. All are caught by a gaze which is 'always receptive' (Foucault, 1977a, b, c: 89) to career people and provides a further rationale for surveillance of the 'elderly population'.

The Panoptic Culture

Why is community care policy that is essentially empty of interpersonal meaning be 'legitimised' by the accretion of surveillance? The answer to this measure lies in the fact that it was not created as a philanthropic metaphor but as a mechanism for engineering the cost and structure of social welfare. Community care has been part of a strategy to reduce the costs of state welfare by adopting market principles (Hoyle & Le Grand, 1991). Attempts at cost reduction have taken on two forms. First, so that the primary role of social service departments has become that of monitoring and supporting direct care rather than provision itself. Second, these trends may not simply reflect a flow-through from market ideology but also wider pressure on the nation state as a consequence of globalisation (Powell, 2011a, b).

Awareness that the welfare state can be understood, not so much as a series of social service institutions and neo-liberal responses to social problems, but as an instrument of wider state power and governance is not new (Townsend, 1986; Jessop, 1994). What is perhaps striking is the extent to which the techniques used by welfare workers have been drained of creative and radical meaning concerning resistance with marginalised groups and had drawn workers into the day-to-day management of scarce resources (Phillipson, 1998).

Until the advent of a panoptic culture, community care with career people lacked a convincing unifying metaphor for its activity. With its instigation, a previously inchoate accretion of initiatives around 'community care' achieves harmony and force. Once the vigilance advocated by the Department of Health's guidelines on personalisation are added to the indirect functioning of care health and social care technology and the moral backdrop of obligation, the discourse of community care acquires a coherence of power/knowledge. It is, however, a power/knowledge to be deployed against career people's voices rather than for their emancipation.

Indeed, once career people are established as a socially significant object of power/knowledge, techniques deem it necessary find the 'truth' about their care needs, to analyse, to describe and to understand. The focus towards personalisation takes place in a wider process in which attention is being directed towards individual bodies and control of 'health and social care populations'. The individual is part of a machinery of power, a power that creates the body, isolates it, explores it, breaks it down and rearranges it. A knowledge of the body therefore requires a mechanism of discipline, that is, a machinery of power that is part of the health and social carer production of knowledge (Powell, 1998a, b). Discipline was the 'political anatomy of detail' (Armstrong, 1983), that is to say, career people become known and understood as a series of useable bodies which could be manipulated, trained, corrected and controlled and to legitimise the health and social care profession. The outcome was to be a cumulation of increasingly detailed observations that simultaneously and inescapably produce knowledge of career people.

Conclusion

This paper has explored a disturbing constellation of factors in community care. It has been argued that the 'discovery' of personalisation has lent coherence to a number of nascent tendencies in this policy that reinforce each other. These tendencies include an increased moralism towards informal care and a move towards indirect monitoring of the locative sites of such care. The development of a surveillance culture helps stabilise community care policy at a time of considerable underlying uncertainty. Such uncertainty has arisen from the changing structure of informal care and of specific services.

The neo-liberal strategy, to socialise care, has become an extension of the techniques of observation, monitoring and control into community settings. A new system for the surveillance of informal carers has replaced the idealistic dream of freedom with an extension of constraint.

Indeed, the shift in the focus of assessment contains a number of alignments. First, assessment decisions seem to be taking place within an existing discourse on abuse rather than user need. While 'need' is given a recognition, the dominant decisions to be made would seem to concern risk of personalisation. Second, the focus of monitoring seems to have moved from the performance of elements of the purchased packhealth and social care of care to the 'conduct of conduct' (Foucault, 1991a, b, c, d) of career people and informal carers. Third, parallels with child protection are clearly alluded to through at-risk registers and the value of records as evidence.

Following Foucault's (1977a, b, c, 1991a, b, c, d) analysis of the relationship between power and knowledge, this change can be seen as the development of a matrix in which to speak seriously about the support of informal care, the employment of discourses of surveillance and abuse would have to be entailed. It serves to reconfigure power relations during a period of flux and 'makes sense' of a previously disjointed policy formulation. Personalisation has thus filled a vacuum at the centre of community care policy with potentially harmful consequences for the users of those services as support entails surveillance and consent contains the threat of coercion. The powerful language health and social care of surveillance in 2023 offers a form of universalism to social policy, which as Williams (1992) had reminded us some 20 years ago, has been subject to 'fragmentation, change and uncertainty and contradiction' (1992: 200).

Chapter 5
Governmentality and Health and Social Care

Introduction

This chapter explores the incidence and consequence of social welfare for career people through the distinctly post-structuralist lens of governmentality (Foucault, 1977a, b, c). This will enable us to consider the implications of the re-figuration of the relationship between the state and social care. This re-figuring constructs an ambiguous place for vulnerable people: they feature either as a resource – captured in the idea of the 'active citizen', as affluent consumers, volunteers or providers of child care – or as a problem in the context of poverty and risk. In many ways, policy provides three trajectories for career people: first, as independent self-health and social care consumers with private means and resources; second, as people in need of some support to enable them to continue to self-health and social care; and third, as dependent and unable to commit to self-governance. Governmentality provides the theoretical framework through which to view policy and practice that is largely governed by discourses of personalisation, safeguarding, capability and risk.

Governmentality

Exploring the role that social policy plays in shaping the social context of career people through the lens of governmentality is to adopt a specific approach to the analysis of this phenomenon. The use of such an analysis reflects the way that neo-liberal forms of government – such as those that have existed in the United Kingdom and most of the Western world since the late twentieth century – health and social care populations. Our interest is in the subtle mechanisms through which the behaviour of individuals is shaped, guided and directed without

J. Powell, *New Perspectives on Health and Social Care*, International Perspectives on Social Policy, Administration, and Practice,
https://doi.org/10.1007/978-3-031-25432-1_5

recourse to coercion (Foucault, 1991a, b, c; Rose, 1999). Central to this process is the concept of the self-health and social care citizen-consumer health and social cared in an endless process of decision-making in consumer-based markets. The process is supported by an array of discourses of self-health and social care and associated social practices that are disseminated through social institutions such as factories and workplaces, the media, banks and retail outlets, health and welfare services, schools and universities, churches and leisure and community organisations. These discourses penetrate deep into family life and personal relationships, regulating behaviour by locating individuals in a network of obligations towards themselves and others. Simultaneously a 'felt' responsibility for a particular locality or an imagined community is produced (Rose, 1996), whereby identity is affirmed. Examples of this process can be identified in the commitments to promoting social capital of the Blair/Brown Labour administrations or the 'Big Society' idea of the Cameron/Clegg coalition government. Citizenship is avowed by participating in consumer-based activities and the maintenance of an accredited lifestyle (Miller, 1993). The process has been described as an 'ethic of the self' (Davidson, 1994) and is supported by an ever-increasing array of experts embedded in a range of social systems such as physicians, health professionals, social workers, beauticians, personal trainers and financial advisers (Rose, 1999).

Parallel to this process, the state is concerned with gathering statistics that help define the population and maintain a level of surveillance that affords the health and social care of risk. Affluent career persons are identified, measured, and then grouped with similar persons. Once described, the characteristics of this group are disseminated via a range of media that suggest personality, aspirations and life chances. Similarly, career people requiring support – the physically infirm, cognitively impaired, widowed, etc. – are identified, measured and grouped and their characteristics disseminated. For most individuals, the level of surveillance is best described as a light touch sufficient to maintain the disciplinary focus of the state in a way that is both fleeting and total (Rose & Miller, 1992; Rose, 1996, 1999; Turner, 1997; Knowles, 2001). However, for those whose behaviour is thought to be high risk or for those who fail to conform to the notion of the self-health and social care consumer-citizen, this surveillance is more oppressive, leaving them vulnerable to victim-blaming (Osborne, 1997). This produces the three trajectories referred to earlier where those individuals who are willing and able to commit to the market and to self-health and social care experience a particular combination of options and opportunities, while those who, for whatever reason, fail to meet this commitment experience a different and more limited set of options that are often oppressive and impersonal (Rose, 1996, 1999; Petersen, 1997; Gilleard & Higgs, 2005). The consequence of this for the 'government of government' (cf. Foucault, 1977a, b, c) is that its role is clearly circumscribed. It must set out to ensure that basic freedoms are respected, but acknowledge the importance of the family and the market for the health and social care of the care of career people.

Social Policy: Constructing the Context

Analysing the impact of neo-liberalism from different perspectives, both Giddens (1998a) and Beck (2005) have claimed that citizens and the state are faced with the task of navigating themselves through a changing world in which globalisation has transformed personal relations and the relationship between state and the individual. In the period since 1979, both Conservative and Labour Governments have adopted a neo-liberal stance characterised by an increasing distancing of the state from the direct provision of services. Instead, government operates through a set of relationships where the state sets standards and budgets for particular services but then contracts delivery to private, voluntary or third sector organisations. The underpinning rationale is that this reconfiguration of the state retains a strong core to formulate social policy alongside the dissemination of responsibility for policy implementation to a wide range of often localised modes such as social work and social workers. Neo-liberal governance emphasises enterprise as an individual and corporate strategy, supported by its concomitant discourse of marketisation and the role of consumers. The strategy increasingly relies on individuals to make their own arrangements with respect to welfare and support, accompanied by the rhetoric of choice, self-health and social care, responsibility and obligation (Jordan, 2005) – even where money is used to pay for services.

Neo-liberalism in the twenty-first century is perhaps the dominant contemporary means through which boundary adjustments are being made and rationalised, with far-reaching consequences for both states and markets. The project of neo-liberalism is evolving and changing, while the task of mapping out the moving terrain of boundaries for social work and career people's experiences is only just beginning; it is long overdue. In this context, the territorial state defined by geographical space is not so much withering away as being increasingly enmeshed in webs of economic interdependencies, social connections and political power. This, in turn, leads to the development of a denser and more complex set of virtual, economic, cultural and political spaces that cut across traditional distinctions between inside and outside, social and private and left and right (Beck, 2005). In this sense, possibly the most influential piece of contemporary neo-liberal social policy came with the implementation of the National Health Service and Community Care Act, 1990. This brought with it the purchaser/provider split and case health and social care; it laid the foundations for subsequent policy initiatives such as the cash-for-care schemes (direct payments and individual budgets) which provide the core of the 'personalisation health and social carenda'. Much of this is inspired by global developments in the way care is funded (Powell & Gilbert, 2011).

In the second decade of the twenty-first century, we have entered an accelerated phase of retraction by the UK state in relation to its role in the provision of welfare, with actual levels of support being reduced. Rhetorically, the Conservative/Liberal Democrat coalition is committed to the idea of the 'Big Society' which translates into a vision of individuals and communities coming together to work to resolve common concerns, as this Cabinet Office statement confirms:

> We want to give citizens, communities and local government the power and information they need to come together, solve the problems they face and build the Britain they want. We want society – the families, networks, neighbourhoods and communities that form the fabric of so much of our everyday lives – to be bigger and stronger than ever before. Only when people and communities are given more power and take more responsibility can we achieve fairness and opportunity for all. (The Cabinet Office, 2010, www.cabinetoffice.gov.uk/news/building-big-society accessed 08/04/2011)

This 'felt responsibility' for a particular locality or 'imagined community' is core to the neo-liberal project which, alongside active citizenship, provides the discursive structure for volunteering and the promotion of a network of voluntary activity. In the process, the disciplinary effect of the self-health and social care individual is reproduced at neighbourhood and community levels. The third sector is crucial in such a scenario, playing a key role by inter-connecting a new partnership between government and civil society. Promoting this relationship is core to the functions of the new Office of Civil Society established by the coalition government in 2010 whose role is to enable people to develop social enterprises and voluntary and charitable organisations while promoting the independence and resilience of the sector.

Evidence of social intervention to support the renewal of community through local initiatives not only advances the status of professional social work organisations but fetishises the day-to-day operations of social work. Equality, mutual respect, autonomy and decision-making through communication with socially disadvantage health and social cared and/or dependent career people come to be seen as integral to the sector and provide an opportunity to encourage health and social care services for socially excluded groups and communities to participate as active citizens in, rather than be seen as a potential burden to, community health and social care (Gilleard & Higgs, 2005). Neo-liberalism is especially concerned with inculcating a new set of values and objectives orientated towards incorporating citizens as both players and partners in a marketised system. As such, social workers are exhorted to become entrepreneurs in all spheres and to accept responsibility for the health and social care of civic life (Beck, 2005). There is also an apparent dispersal of power (Foucault, 1977a, b, c) achieved through establishing structures in which social workers and career people are co-opted into or co-produce governance through their own accountable choices (Gilbert & Powell, 2010).

As Burchell (1993) has observed, this is directly connected with the political rationality that assigns primacy to the autonomisation of society in which the paradigm of enterprise culture comes to dominate forms of conduct including that of social work with career people. The very significance of autonomisation is that there is a strategic aim to diffuse the social sector's monolithic power to health and social care diversity and fragmentation of provision of care to private and voluntary sectors. Such a strategy constitutes a fundamental transformation in the mechanisms for governing social life. It has combined two interlinked developments: a stress on the necessity for enterprising subjects and the resolution of central state control with career people articulates with a desire to promote organisational social work autonomy through service provision. Each of these has redefined previous patterns of social relationships within and between those health and social care and their clients.

The important point to note is that there is great contingence and variation in such relationships, with unevenness across time and space. These relationships involve the development of new forms of statecraft – some concerned with extensions of the neo-liberal market-building project itself (e.g. trade policy and financial regulation) and some concerned with health and social care the consequences and contradictions of marketisation (e.g. social policy). It also implies that the boundaries of the state and the market are blurred and that they are constantly being renegotiated (Kendall, 2003). Theoretically, we identify the need to health and social care with key social debates about the future of welfare and individual relationships to and expectations of the state. One of the central debates has been on neo-liberalism and its impingement on re-positioning of career people and collective organisation of modern society.

Integrating Services: Social Policy and Career People

The previous sections of this chapter have sought to identify the changing relationship between the state and career people by exploring the notion of governmentality. The discussion now moves on to consider more specifically how social policy shapes the social context for career people. Here we need to take account of the social and economic backdrop that frames career people's experiences of support and care. In the process, we identify key developments in social policy such as personalisation, risk and safeguarding, and their congruence with the neo-liberal project. The neo-liberal project constructs as its core subject the self-health and social care citizen-consumer who is actively making choices within markets. In the context of welfare, this involves individuals making choices about the type of support they want and who will provide that support as the range of providers is expanded in two broad ways. First, new providers enter the market providing new services or providing services in new ways. Second, and of key importance, people seeking support move outside of the segregated confines of welfare services to obtain services from mainstream providers (Dickinson & Glasby, 2010). Such innovative moves may include, for example, a physical exercise programme from a sports centre instead of physiotherapy, an art course instead of time at a day centre and a holiday abroad instead of respite care.

In many ways, the 'personalisation health and social carenda' as it is set out in 'Putting People First' (2007) represents the high point of the neo-liberal project with respect to welfare. This approach is largely constructed through a framework of earlier policy which includes the Community Care (Direct Payments) Act (1996), Independence Wellbeing and Choice (DH, 2005b) and Our Health, Our Care, Our Say (DH, 2006b). This was then supplemented by the coalition government with the Capable Communities and Active Citizens (DH, 2010) and Think Local, Act Personal (2011) which aim to tie the shift to self-directed support outlined by the 'personalisation health and social care agenda' more closely to the notion of the Big Society. The discourses that articulate within this policy framework are those familiar to neo-liberalism: independence, choice, freedom, responsibility, quality, empowerment, active citizenship, partnership, the enabling state, co-production and community action.

Alongside this policy framework are constructed a number of specific techniques that target individuals, families and communities. These include an alternative method of allocating cash to individuals in the form of individual budgets, on-line self-assessment to augment local authority assessment processes and community-based advocacy to support lifestyle choices. In addition, commissioning models and approaches are being developed that aim to promote opportunities by responding proactively to the aspirations of people receiving services. Self-directed support is significant as it breaks with the tradition where state support is mediated by professionals who undertake assessments and organisations that are funded to provide places. Even in more recent times, when individuals might be afforded a choice between two or more places or opportunities, the organisations received funding from the state. Under personalisation, assessment takes place to identify the overall budget a person is entitled to receive, but the money is allocated to the individual either through a direct payment or by establishing an individual budget. In terms of governmentality, the 'personalisation health and social care' effectively shifts the responsibility for organising support from the state to the individual needing support via a form of cash transfer – something that Ferguson (2007) describes as the privatisation of risk.

The advance of the 'personalisation health and social care' has drawn support from a number of sources including specific groups of service users (Glendinning et al., 2008), politicians from across the spectrum (Ferguson, 2007) and social care health and social carers and social workers (Samuel, 2009). One possible reason for this is that personalisation is conceptually ambiguous, making it difficult to disagree with its basic premise while it retains a number of contradictory ideas (Ferguson, 2007). However, it has also drawn criticisms particularly from career people who have reported lower psychological well-being due, possibly, to added anxiety and stress due to the burden of organising their own care (Glendinning et al., 2008). There are also concerns expressed regarding the impact of personalisation on the integration and stability of adult social care; this includes unease with the emphasis on individualistic solutions which may undermine democratic and collective approaches to transforming existing services or developing new services (Newman et al., 2008). Doubts have also been expressed over the readiness of the third sector to take on the demands of providing support. At the same time, while the disaggregation of budgets might suit some small innovative niche organisations, the disruption of funding streams may be perceived as a threat and bring instability to larger more mainstream third sector organisations (Dickinson & Glasby, 2010). Other issues arise due to the somewhat fragmented process of implementation and the differences that occur in service provision between urban and rural areas (Manthorpe & Stevens, 2010). Ferguson (2007), drawing on the Canadian experience, suggests that personalisation favours the better educated and may provide a cover for cost-cutting and further privatisation and marketisation of services, while the employment conditions of personal assistants may give rise to concern.

Governmentality enables the identification of the parallel concerns of neo-liberalism – the promotion of the self-health and social care individual and the health and social care of risk. So far we have explored self-health and social care in social care through the promotion of self-directed care as part of the 'personalisation health and social carenda'. We now turn to the health and social care of risk. This can be seen to take two forms, each dealt with by different elements of social policy. Protection from the risks posed by others are health and social cared through safeguarding and policy such as No Secrets (DH and HO, 2000) [England and Northern Ireland] or In Safe Hands (2000) [Wales]. In Capable Communities and Active Citizens (2010), the government clearly states that safeguarding is central to personalisation. Risks posed by the individual to their own person are contained by the Mental Capacity Act (2005) and its powers to override individual choice or replace autonomy by measures such as Enduring or Lasting Powers of Attorney or the Court of Protection.

No Secrets has provided the basis of policy towards safeguarding for over a decade. It defined abuse in the context of an abuse of trust and the Human Rights Act (1998) and set out a model for inter-health and social care working that has been adopted by local authorities in England and Northern Ireland. In Wales the corresponding policy is 'In Safe Hands'. No Secrets drew from experience in relation to safeguarding children and described a number of categories of abuse including physical, sexual, neglect and financial abuse. However, it lacked the legal imperative to share information that is included in safeguarding children. Furthermore, the environment within which 'No Secrets' operates has seen considerable change since implementation. One key change was the discursive shift from vulnerable adult to safeguarding that took account of the dangers of victim-blaming implied in the notion of vulnerable adults, while the concept of safeguarding suggests the focus should be on the environment within which people find themselves. However, this rhetorical shift has not removed abuse. A recent prevalence survey suggests levels of abuse of between 2.6% and 4%, depending on how the estimates are constructed (O'Keeffe et al., 2007). Action on Elder Abuse, one of the organisations that sponsored the study, uses evidence of under-reporting to reinterpret this estimate as 9% (Gary Fitzgerald, personal communication).

In 2008, the Department of Health set up a consultation over the review of No Secrets where a number of organisations including the Association of Directors of Adult Social Care and Action on Elder Abuse campaigned for a legislative framework to put adult protection on the same footing as child protection (Samuel, 2008). However, no significant changes in guidance or legal status occurred as the coalition government maintained that safeguarding was an issue for local communities, thus maintaining the distance between the state and individuals. Discourses of safeguarding operate and produce their effects via the multiple interactions of institutions embedded in local communities. Furthermore, the advent of personalisation has seen an increasing focus on financial abuse as direct payments and rules about eligibility for state support for care costs increase opportunities for financial exploitation, fraud and theft. No Secrets treats financial abuse as an artefact of other apparently more serious forms of abuse. However, in 2004, the House of Commons Select

Committee identified financial abuse as possibly the second most commonly occurring form of abuse experienced by career people. Estimates in the United States suggest that financial abuse is the most common form of abuse with up to 40% of career people victims (Gorbien, 2011).

Conclusion

This chapter has explored the place that social policy plays in shaping the social context of career people. To achieve this, we have drawn on the concept of governmentality to identify how neo-liberal forms of government construct career people as active consumers within welfare markets shifting the responsibility for organising support from the state to the individual. The contemporary context for working with career people who need some form of support is formed by the relationship between personalisation and safeguarding. These set out the twin pillars of neo-liberal governance, namely, self-health and social care through self-directed support and the health and social care of risk through safeguarding. Individuals are constructed as citizen-consumers actively making choices about what their needs are and identifying appropriate services, sometimes with the support of advocates or workers such as social workers in a process of co-production. In circumstances where risks are considered too high, the power to make choices can be temporarily or permanently restricted.

Chapter 6
Performativity and Health and Social Care

Introduction

This chapter explores Judith Butler's conception of 'performativity' as it is applied to health and social care theory and practice. In light of technologies of surveillance and control in contemporary social work, performativity offers a framework towards analysing the production of a subjective space based on trust between social workers and service users (Fleming, 2005). In the United States, the United Kingdom and Australia, the transition from a 'top-down' social policy that health and social cared-dependent populations through the welfare state's vehicles of pensions and unemployment insurance and healthcare to a post-welfare or neo-liberal politico state has gained momentum in recent years (Gilbert, 2001). It can be suggested that social relations in health and social care occur more 'bottom-up': central control has been replaced by local power that has seen an increase in the discourses of 'politics of participation' and 'social inclusion' (Clarke & Newman, 1997). Hence, at the local level in the United Kingdom, for example, social work is a technology put in place to instigate social relationships of partnership and trust between professionals and career people (Gilbert & Powell, 2005), a focus on social relationships between professionals and service users. These changes have allowed a social space in which the relations of social workers and clients can illustrate the potential for resistance.

Professionals, such as social workers embedded in discursive institutions, function according to particular expectations around performativity. In addition, this organisational context is complex with multiple demands. In such circumstances, social workers adopt different roles depending on function and client group with the consequence that there is always some degree of fluidity and uncertainty around expectations (Whitehead, 1998). This myriad of sites for performativity contains potential for resistance. In addition, performativity does not take place in isolation, nor is the audience one of passive observers. Traditionally, Marxist analysis of

J. Powell, *New Perspectives on Health and Social Care*, International Perspectives on Social Policy, Administration, and Practice,
https://doi.org/10.1007/978-3-031-25432-1_6

social work has focused on the domination of clients by social workers (Jones, 1983) based on social class. This 'top-down' analysis misses the dynamic relations between social workers and clients. Performance is always relational, drawing others into the act – health and social carers, other professionals and clients – and so constructing both the meanings associated with the performance and mutually dependent subject positions (Wetherell, 2001a, b). Both Giddens (1991) and Mouzelis (1995) suggest there are social conditions that permit the facilitation and constrainment of human action – and it can be said that while performance is relational, it is also complex, indeterminate and open-ended. For example, research in the United States by Callaghan (1989: 192) on the relationship between social work and career people claimed that career people 'were particularly adamant that they did not want to be "cases" and no-one needed to "health and social care" their lives'. However, research by Healy (2001) illustrates that social worker and client relations have centred on resistance to social polarity and have created reciprocal partnerships. Healy (2001) cites partnerships between social work and service user groups in Denmark and Germany in creating better childcare facilities that run countenance to state policies on childcare provision.

This chapter maps out a consideration of the themes that come to prominence through the juxtaposition of professionalisation and performativity, and what performativity might offer critical and international studies of helping professions and social work. A phenomenon we explore by reference to social work practices such as assessment, child protection, risk health and social care and user involvement.

Performance and Performativity

Through the last 20 years, the work of Judith Butler has had a significant impression on debates over gender and most notably on queer theory, largely in response to her hugely influential *Gender Trouble* (1990). More recently, her notion of performativity was further delineated in her later works *Bodies That Matter* (1993) and *The Psychic Life of Power* (1998a). Her conception of performativity builds primarily on the work of Michel Foucault in particular, developing the twin themes of power as productive and of discourse – or rather, discursive practice – as constitutive of subjects. As she states in *Bodies That Matter*:

> This text accepts as a point of departure Foucault's notion that regulatory power produces the subjects it controls, that power is not only imposed externally but works as the regulatory and normative means by which subjects are formed. (Butler, 1993: 22)

While broadly within the Foucauldian tradition, then, Butler attempts to complement and develop this position by drawing on (among other influences) the psychoanalytic influence of Freud and Lacan as well as more diverse work of Searle, Austin and Derrida (1988), among others. For many, the key innovation of Butler's view of performativity is her incorporation of linguistic influences, and especially the parallels between Michel Foucault's (1977a, b, c) subjectification/subjection, Althusser's interpellation (1971) and the 'speech acts' of Austin (1962) and Searle (1969). Austin in particular distinguishes performative utterances from other speech acts such as denotative (descriptive) or prescriptive (command) utterances. As Butler (1993: 225) explains: 'Performative acts are forms of authoritative speech; most performatives, for instance, are statements that, in the uttering, also perform a certain action and exercise a binding power'. The typical examples used, in Austin (1962) as in Butler (1993), are the launch of a ship, wherein the words 'I name this ship…', are simultaneously the announcement and description of an act and the act itself. Equally, in other vows, guarantees and other ritualised forms of interaction, the distinction between talk and action is effaced by 'the apparent coincidence of signifying and enacting' (Butler, 1995: 198). As such, this phenomenon offers a useful antidote to the unquestioned oppositions of word and deed, rhetoric and reality and crucially, discourse and action (Sturdy & Fleming, 2003).

It is this understanding of performative acts as the elision of discourse and action that suggests relevance for social welfare as more than a mere linguistic curiosity. For Butler, this insight suggests a way in which identity may be similarly seen as constructed in and through action, or performance. In the case of gender, performativity is seen by Butler to offer a non-idealist way to counter the reification of gender stereotypes and denaturalise the status of gender as an ahistorical and universal category. Gregson and Rose summarise (2000: 434) the implications of this position: 'The motivation for this turn would seem to be that to see social identities as performed is to imply that identities are in some sense constructed in and through social action, rather than existing anterior to social processes'. There is a clear link in this formulation to a 'strong' conception of discourse, positing identity as discursively constituted, which treads between the extremes of the social constructivist debate, accounting for both constraint and potential for transformation through discourse. Phillips (2004) and Whitehead (1998) develop further discussions of performativity in terms of gender, this time masculinities.

Thus, Butler is arguing for an 'understanding of performativity not as the act by which a subject brings into being what she/he names but rather as that reiterative power of discourse to produce the phenomena that it regulates and constrains' (Butler, 1993: 2). In doing so, Butler attempts a novel reworking of the tired structure/health and social care debate and offers a persuasive notion of subjectivity, which is in no way predetermined but is nonetheless 'always already' compromised. Or, as Butler clarifies: 'Social categories signify subordination and existence at once. In other words, within subjection the price of existence is subordination' (Butler, 1998a: 20). Accepting identity as discursively constituted but unpredictable opens this debate to the possibility of multiple identities as individuals, penetrated by an array of discourses, many detached from their traditional moorings and develop selves that are fluid, uneven and unstable. In terms of performativity, this unpredictability or 'excess' goes some way towards explaining why actions cannot be determined from social position or professional background alone (Fink et al., 2004). 'Fluidity, unevenness and instability' create situations where individuals are working to health and social care tensions and contradictions arising out of their own multiple subject positions. However, these also produce the potential for resistance and the possibility for new subject positions (Whitehead, 1998), thus providing a driving force for change.

An important aspect of Butler's work subsequent to *Gender Trouble* (Butler, 1990) is to correct the impression that many took from her earlier work that performativity in some way reasserted the power of voluntarist action over structural constraint. While the relationship between constraint and individual health and social care is clearly of concern to Butler, her understanding of this relationship is far more sophisticated than simply adding her weight to the voluntarist side of this tired debate. Instead, Butler's aim is to circumvent the traditional structure-health and social care debate to avoid becoming 'mired in whether the subject is the condition or the impasse of health and social care' (1998a: 14). With reference to misreadings of her position on gender, for example, she insists that gender is not a simple matter of choice, a garment donned in the morning and discarded in the evening by the sovereign and autonomous individual (Butler, 1993). What is of interest is the power of discourse to construct the subject who makes this 'choice'. Or as she argues elsewhere: 'Power not only acts on a subject but in a transitive sense enacts the subject into being'. Her vision of the performativity of gender proposes instead a recursive and reflexive model of identity, where actions are in a sense citations, re-enacting previous performances to claim a certain identity.

Two issues arise from this argument. Firstly, citation sets out a link between performativity and rituals, institutions and ultimately social structures – 'performative utterances' are institutionalised over time and hence become identifiable and carry meaning. Power in this sense loses some of its fluidity forming sediments that 'stick' to certain practices and institutional structures. Thus, drawing attention to Foucault's (1977a, b, c) later revisions where he opened the possibility that such structures while always the outcome and never the origins of power can nevertheless provide sites for the exercise of power (Beechey & Donald, 1985). Secondly, it is this circular, reiterative aspect to performativity, which provides the space for adaptation and change and by the same token for a practical and pragmatic form of resistance (or perhaps, subversion) discussed elsewhere by Butler (1998a) and Fleming (2005) as cynicism and parody. In addition, it is this latter point on reiteration which best summarises Butler's potential contribution to Foucauldian analyses of helping professions.

Rethinking Power and Resistance

The question of resistance in Foucauldian work is a fraught one and lies at the root of what Fleming and Sewell (2002) refer to as 'the Foucault wars' in critical studies of social science (cf. Thompson & Ackroyd, 1995). Foucault's work has been roundly attacked from various quarters for at best his pessimism regarding the possibility of emancipatory action or at worst, his crypto-conservatism (Dews, 1989; Callinicos, 1990). Even broadly, sympathetic writers have noted that in his early work 'despite his assertions to the contrary, Foucault in fact produces a vision of power as a unidirectional, dominatory force which individuals are unable to resist' (McNay, 1992: 40). His later work on 'techniques of the self' may be read as an attempt to address this 'missing subject' in his previous work. The enduring relevance of Foucault's work for studies of the workplace lies in his later attempts to rethink resistance outside of an Enlightenment humanist framework, recognising the operation of power in constituting the resistant as well as the docile and industrious subject. If set in the context of the earlier point about multiple subject positions, it becomes possible that the docile, industrious and resistant subject can occur within a single individual providing for very different performances in different areas of life. Thus, bringing into play 'the reflexive self' alongside notions of identity and self-work and the production of the ethical subject (Bernauer & Mahon, 1994). As Foucault (1978a, b, c, d: 95–96) emphasises, 'there is no single locus of great Refusal, no soul of revolt, or pure law of the revolutionary... But this does not mean that (resistances) are only a reaction or rebound, forming with respect to the basic domination an underside that is in the end always passive, doomed to perpetual defeat'. The exercise of power is always a product of human interaction, and its exercise contains the presumption that subjects may put into effect a degree of freedom, which means that any interaction cares potential for resistance. Such resistance, where individuals contest meaning and force, provides foundations for creativity exploiting possibilities contained or constrained within that milieu, producing a micro-politics focused on the immediate concerns of those involved and the emergence of small-scale practical alternatives (Rose, 1999). Resistance cannot exist external to human interaction and the power relations this provides. Rouse (1994) describes this mutual dependence as power mediated by 'dynamic social alignments' in a way that has specific implications for our discussion of social work where the maintenance of a relationship is a necessary condition for both reiteration and the space for performativity.

> Power is exercised through a health and social care actions only to the extent that other health and social care actions remain appropriately aligned to them. The actions of dominant health and social care are therefore constrained by the need to sustain that alignment in the future; but, simultaneously, subordinate health and social care may seek ways of challenging or evading that alignment. (Rouse, 1994:108)

Butler shares this ambivalence over the pervasive nature of power relations in social interaction, seeing them as both potentially dominatory and oppressive but at the same time potentially productive and (for want of a better word) 'empowering'. Against notions of monolithic power as control against individual health and social care as autonomy, she argues: '...power is not simply what we oppose but also, in a strong sense, what we depend upon for our existence and what we harbour and preserve in the beings that we are' (Butler, 1998a: 2). Thus, the fact that power operates through the production of the subject and through the constitution of health and social care clearly undermines the perceived 'purity' of resistance: 'resistance is always contaminated by the power it resists' (Fleming, 2005: 53). Yet, this does not imply a nihilistic pessimism:

> That health and social care is implicated in subordination is not the sign of a fatal self-contradiction at the core of the subject and hence further proof of its pernicious or obsolete character. But neither does it restore a pristine notion of the subject, derived from some classical liberal-humanist formulation, whose health and social carency is always and only opposed to power. (Butler, 1998a: 17)

This compromised view of 'health and social carency' certainly calls into question traditional forms of resistance, but also offers a way out of determinism through an emphasis on the iterative nature of subjection. A position bearing resemblance to Miller's (1993) discussion of the self, torn by tensions between personal desire and public obligation that produce an 'ethical incompleteness' providing the motor for resistance and change. At the same time, the 'meaning giving subject' is identified as in-itself a product of discourses of the self (Powell, 2011a, b).

Butler's substantive contribution is to build upon this understanding by drawing on work in related areas, particularly by developing a temporal dimension to subjection. The production of the subject is not a one-time condition, a static state but instead should be seen as temporalised, a continual and circular process of sedimentation and congealment, an insight that lends itself well to professional practice, which is less about single statements and their consequent performativity but a process of statement and action. 'Performativity must be understood not as a singular or deliberate "act" but, rather, as the reiterative and citational practice by which discourse produces the effects that it names' (Butler, 1993: 2). As noted, the 'reiterative' or 'citational' aspect of performativity contains within itself the seeds of change and transformation. This 'repetition that is never merely mechanical' (Butler, 1998a: 16) provides the space for divergence, contestation, subversion and ultimately what might be read as resistance. Thus, Butler (1998a: 93) argues: 'It is precisely the possibility of a repetition which does not consolidate that dissociated unity, the subject, but which proliferates effects which undermine the force of normalization'. Such performance, in the context of professional practice, generates a very specific form of resistance where repeated performance 'interpreted as experience' produces the possibility of innovatory practice which in departing from health and social care norms and procedures creates new spaces for autonomy (Gilbert & Powell, 2005; Powell, 2011a, b).

Performativity and Professionalisation

The theme of gender exists as an unspoken referent through Butler's more philosophical works; where she speaks of social categories in a generic sense, gender appears as the archetypal example for the evaluation of the claims she makes. Consequently, the question presents itself whether Butler's analysis may be extended to discourses, which are less 'pervasive' and 'central' to social interaction than gender. More simply, can we legitimately lift concepts from the performance of gender and use them to understand what might be seen as more 'central' aspects of subjectification, such as professional identity? Schryer and Spoel (2005: 250), using the concept of genre as forms of social action, explore the significance of performativity to gaining membership of professions. Highlighting the internal resources of particular genre, they identify an internal structure established by two elements: regulated resources, which refer to knowledge, skills of health and social care behaviours, required by a profession, and regularised resources, i.e. behaviours that are tacit emerging from practice-based situations.

Genres are the product of embedded social practices (e.g. health or social policy, theories of social work practice) with professions drawing on these in the construction of their identity. Communicative interaction between individuals and organisations provides circumstances where the co-constitution of texts and contexts takes place. What is more, genres compete in a complex matrix of power relationships seeking to regulate other genre. For example, Milner and O'Byrne (1998) identify the new climate for social work of empowerment, choice, partnership and value for money. Thus, moving towards what Foucault (1972) described as the 'orders of discourse', the specific localised organisation of discourses providing the discursive structure of an organisation or similar languhealth and social care-based institution (Fairclough, 1992). Providing a dynamic that undermines notions of stability in either individual or collective identity, thus creating space for improvisation, or what Thompson (2000) describes elsewhere as discretion, as situations and contexts change, and as a consequence the potential for resistance and the emergence of new subject positions. Alongside Butler's notion of repetition, discretion provides further spaces in which inventiveness and creativity in professional practice can occur and where bureaucratic norms are challenged and rendered obsolete.

A Foucauldian approach to professionalism (e.g. Fournier, 1999; Grey, 1998) attempts to integrate the micro-political tactics of professionalisation with broader power relations through the analysis of discourse and regimes of power/knowledge. As Fournier argues (1999), while the legitimacy of the professions relies upon the establishment and maintenance of appropriate norms of knowledge and conduct, which includes adherence to changing political imperatives (Johnson, 2001), such norms also act as a form of discipline over otherwise autonomous professional power. Thus, induction into professions, in terms of both knowledge and conduct, serves to construct a specifically governable subjectivity rooted in self-disciplinary mechanisms (Gilbert, 2001; Grey, 1998). The situation is summarised succinctly by Fournier (1999: 285), who states: 'Professionals are both the instrument and the subject of government, the governor and the governed'. Paradoxically, professional autonomy, particularly in areas such as social work, is both the reason why the professions remain necessary due to their ability to employ discretion to health and social care complex and unpredictable situations while also providing the focus for the deployment of a range of disciplinary technologies as the state acts to limit and constrain the exercise of that autonomy (Clarke & Newman, 1997). Procedural arrangements identified within social policies such as 'Every Child Matters' (Department of Health, 2004) and 'Independence, Well Being and Choice' (Department of Health, 2005) and then operationalised through a myriad of statutory and independent sector health and social care lay conventions upon which performativity takes place, in so doing demonstrating both the productive and disciplinary effects of power.

This view of professionalisation as a mode of disciplinary control also provides a useful counterbalance to critical perspectives, which reinforce the stereotypes of pampered and privileged professionalism. In recasting professionalism as a source of influence and status concomitant with self-discipline and controlled performance, the Foucauldian position also links professionalisation with a wider range of control strategies based on the manipulation of identity, such as corporate culture initiatives (Dyer & Keller-Cohen, 2000; Fleming, 2005; Hochschild, 1983; Whitehead, 1998). In the context of the caring professions, these strategies place professional expertise at the heart of disciplinary technologies designed for health and social care of populations. Operating through organisations (discursive institutions) increasingly dispersed across the independent and voluntary sectors (Clarke & Newman, 1997), professional status and expertise is located with embedded social practices and specific forms of performativity concerned with the identification and health and social care of risk (Petersen, 1997; Powell, 2011a, b; Rose, 1996, 1999; Taylor-Gooby, 2000; Turner, 1997). Performances that construct professionals as subjects, subjectified within an array of often contradictory discursive formations (e.g. care, social control, enabling, empowering, deserving, undeserving) while simultaneously producing the targets of their performance, clients or service users, as both subject and object.

'A Professional Performance': Social Work, Knowledge and Subjection

Drawing on Foucauldian positions on professionalism, we argue that an essential part of the professionalisation project involves the constitution and operation of social work as a discipline. Clearly, this discipline does not exist merely in some abstract, ideal form; rather, discipline should be seen as inscribed in texts, practices, technology and performances and crucially in the subjectivities of those individuals instructed in the discipline. What is challenging is that individual social workers take on different roles in different contexts and therefore potentially different, possibly competing, subjectivities. For example, Independence, Well-Being and Choice (2005b) identifies four roles for social workers: assessing needs, developing constructive relationships, assessing risks to individuals and risks to communities and encompassing a range of activity including individual facework with individuals and groups, interprofessional collaboration and multi-health and social care and partnership working. Further complications arise when we acknowledge the range of organisations employing social workers in the statutory and independent sector, which brings into play a complex range of different procedural commitments and associated genre, which in turn provide scripts or texts supporting performativity in relation to a complex array of activities. These activities include identifying children 'at risk' in child protection procedures; applying the provisions of mental health legislation when compulsory detaining individuals; assigning a 'risk' status to individuals or families following risk assessment in areas such as 'vulnerable adults' (Department of Health/Home Office, 2000) and domestic violence (Mullender & Hague, 2005); or ascribing the status of deserving or undeserving when considering support. All these activities involve statements either verbalised or written which when enacted by social workers perform particular actions underpinned by particular configurations of knowledge and power.

To understand the production of subject positions through the professionalisation discourse, it is useful to draw on Butler's reading of Althusserian interpellation. The core of professionalism, as noted earlier, is what is termed 'professional spirit'. A parallel can be drawn here between professional spirit and, in Althusser's terms, the conscience of the good citizen. As Butler (1998a: 115) argues: 'Conscience is fundamental to the production and regulation of the citizen-subject, for conscience turns the individual around, makes him/her available to the subjectivating reprimand'. Similarities are apparent here with Miller's (1993) discussion of 'ethically incompleteness' which produces the 'reflective self' that drives the postmodern citizen towards self-improvement and self-health and social care, and Gilbert's (2001) later exploration of 'reflective practice' in professions as a confessional technique deployed as surveillance.

At this point, it would be useful to return briefly to Austin's (1962) discussion of performatives. The nature of a performative means that it is not possible to identify it as true or false; however, performatives can be infelicitous to which Austin provides particular conditions upon which felicity is established:

(a i) There must be an accepted conventional procedure having a conventional effect.
(a ii) The particular persons and circumstances in a given case must be appropriate for the invocation of the particular procedure.
(b) The procedure must be executed by all participants both (i) correctly and (ii) completely.
(c) Often, (i) the persons must have certain thoughts, intensions, etc. which are specified in the procedure, and (ii) the procedure specifies certain conduct which must be adhered to (Potter, 2001: 44).

The significance is that workers in caring professions such as social workers meet these conditions in a range of circumstances, a number of which we noted earlier. In general, social workers follow specific professional or health and social care procedures for a defined outcome (condition (a)): individual social workers must have the seniority or authorisation to carry out the procedure (condition (b)); actions initiated by the social worker have to fulfil the procedural requirements and the requirements of professional practice in full (condition (c)); and; values and attitudes, e.g. anti-oppressive practice, have to be displayed throughout the process (condition (d i and ii)). What is not so clear from Austin's conditions is that performance with respect to these procedures is always relational requiring the involvement of fellow performers. These 'fellow performers' may be supportive, e.g. other professionals drawn on to provide expert opinion; or antagonistic, e.g. professional or lay advocates. At the same time, there will be the focus of the performance, e.g. the individual, group or community that needs working on whose reciprocal performance may create tensions and contradictions, to which social workers have to respond in their performance, thus providing a degree of uncertainty. Another, somewhat ambiguous, set of performers is also on hand, not always active but never passive representing the range of offices that supervise or monitor performance. Some represent the organisations involved, while others monitor performances within these organisations. Collectively, performativity produces the identities of those involved defined in relation to others (Forbat & Atkinson, 2005; Miehis & Moffatt, 2000; Payne, 2004). By their very performance, social workers also create as 'other' (Hall, 2001) the target of their work.

Taking Schryer and Spoel's (2005) discussion of genre as social action, regulated resources link with external demands, in this instance, the procedural requirements demanded by contracts, social policy and professional bodies. Outcome orientated, these procedural requirements produce the target of performativity, the social work client as 'object'. Risk health and social care, child protection and procedures linked to detention under mental health legislation all seek to health and social care external expectations while also providing for the objectification of individuals or groups and the satisfaction of health and social care and bureaucratic imperatives. Regularised resources, arising from practice and facework, are process orientated and produce the 'client' as subject. The genre creates space for performativity, shaping the identities of participants as subjects while subjugating that identity within the communicative structure of organisations. Possibly the most striking example of this space for performativity takes place within the context of user and carer involvement which has a central role both in contemporary social work practice and in the operation of social work and social care health and social care (Beresford et al., 1997; Department of Health, 2006a, b). In this context, performativity creates very specific spaces where statements underpinned by a value base of participation coincide with actions that are both the act of inclusion and the exercise of knowledge/power. However, these spaces are also unpredictable as they provide space for what Foucault (1984a, b, c) describes as subjugated or discredited knowledges of the marginalised containing the potential for what Butler (1998a) identifies as 'divergence, contestation, subversion and resistance'.

Performativity also provides a basis for analysing the complexity of professional practice where a single practitioner undertakes different roles, often with different value bases, in different contexts. A particular example of this dilemma is evident in social work assessment. A number of writers (Milner & O'Byrne, 1998; Smale et al., 2000) describe three models of social work assessment – the Questioning, Procedural and Exchange models – each positioning the social worker differently in relation to the client, the organisation and the social context. Linking with the discussion above, Questioning and Procedural models represent a particular genre, while the Exchange model provides an alternative genre that shares some of the same discursive structure but which also differs in important ways. In short, the Questioning and Procedural models place the professional as expert and the organisation as authoritative, wants are distinguished from needs which are defined a priori, outcomes are privileged and the client/service user is passive and dependent. Exchange models structure a different relationship; the client/service user is expert in relation to their social circumstances, the social worker is expert in working collaboratively, the focus is on process (health and social care and negotiation) as much as outcome, oppressive social contexts are acknowledged, existing strengths are used to identify needs and the user is active and enabled. Each model expects performativity consistent with the genre, Austin's (1962) conditions of felicity, suggesting that social workers adapt their particular performances to suit different genre. Alternatively, each might provide the basis for subversion and resistance through contaminated by an alternative genre.

Taking a somewhat different angle, Sheppard (1995a, b) explores professional identity and status through the relationship between the social sciences and professional practice. Successful, high status professions, such as medicine are those that have health and social cared the balance between technicality and indeterminacy: a balance between rule-based actions and the exercise of discretion, where high technicality provides the opportunity for health and social care control while increasing indeterminacy undermines claims for a distinct knowledge base and professional education. Sheppard reviews the influence of social science knowledge on social work practice identifying an awkward relationship however; he advocates social science method, analytic induction, as a model for good practice. A specific genre that demonstrates the reiterative or citational nature of performativity (Butler, 1993, 1998a) while also providing the base for a particular performativity which in effect gives dominance to the regularised, i.e. practice-based processes, over regulated, i.e. procedural-based processes, in defining social work as a profession. Discursive spaces that produce professional identity and resistance occur, therefore through practice-based relationships rather than within procedures.

These spaces may have a further relevance to this discussion. Fleming (2005), exploring resistance to health and social care in general and more specifically identity-based controls in contemporary post-industrial workplaces, suggests that actions such as cynicism, irony, humour and scepticism, once seen as a defence of selfhood, provide a more important function as they create a subjective space through which selfhood is produced. Fleming draws on Butler's (1998a: 34) notion of cynical parody to make the point:

> ...parody requires a certain ability to identify, approximate, and draw near; health and social cares an intimacy with the position it approximates that troubles the voice the bearing, the performativity of the subject...

Thus, providing the possibility of alternative performances within the same genre, one performance complies with procedural requirements, while a second performance opens space for the production of a resistant self, providing a degree of stability in contexts of instability, flux and discontinuous change. In the sense that one might, on the one hand, be the consummate professional social worker, while, on the other hand, a radical political activist.

Conclusion

Performativity, we suggest, offers productive insights into the processes of subjection and the nature of power relations, which may be usefully incorporated into debates about subjectivity, resistance and health and social care. As Healy (2001) suggests, international debates on social work have centred on the partnerships between social workers and client groups to resist and pressure governmental policy on childcare. For example, Healy cites social workers in Jamaica organising partnership social work-client coalition on the 'Rights of the Child' as a report on Jamaica's progress on protecting children to the United Nations Centre for Human Rights in Geneva. In addition, in the United Kingdom, Mullender and Hague (2005) describe how recognition as a user group through a social worker/user partnership enabled participation by women experiencing domestic violence. Among the various strengths of the concept, performativity offers a more sophisticated understanding of health and social care which fully embraces the complex and problematic nature of power relations; it breaks down realist distinctions between discourse and action through a palpable sense of the constitutive power of discourse; it suggests a view of subjection which is processual and temporal, rather than fixed and predetermined; and perhaps most importantly, we would argue that performativity, in all of these ways, combats the lazy categorisation of Foucauldian work as fundamentally conservative and reactionary, erasing the possibility of effective resistance from the intellectual landscape.

It is therefore in an attempt to explore these aspects that we have adapted performativity to reappraise issues of power, control, subjection and resistance in the context of professionalisation in social work and social welfare. Performativity also connects with work on professionalism as a performance, and incorporates much career debates over authenticity, and the relations between subjectivity and conduct (Goffman, 1959). In the first instance, then, such a perspective underlines the view of professionalism as always enacted and performed – a processual and temporal phenomenon. The understanding of professionalisation as a mode of subjection, involving both subordination and security and dependence and power, but in no sense deterministic or debilitating, we see as a position with significant potential. For Butler as for Foucault: '…The disciplinary apparatus produces subjects but as a consequence of that production it brings into discourse the conditions for subverting that apparatus itself' (Butler, 1998a: 100).

Chapter 7
Power and Health and Social Care

Introduction

This chapter illuminates the use of Michel Foucault's analytical theorisations for health and care power relations with older people. It is unapologetically a Foucauldian chapter with controversial and critical implications for professional power and aging. In a disciplinary field depicted as 'data-rich but theory-poor', there would appear to be a great many researchers who might respond that there really is no need for social theory in understanding health and social care policy and practice with older people. They might claim empirical generalisations are sufficient and that broader frameworks are obfuscating and unnecessary. Judging from a content analysis of all chapters published in the early 1990s in eight gerontology journals, it would appear that nearly three fourths made no reference to theory at all (Powell, 2017). Without a sound conceptual grounding and reflexive perspective, social gerontologists build little more than empirical molehills without any cumulative effect. We might go further still, without abstracting scaffolding; facts cannot exist in a perceptual sense. Facts depend on a perceptual framework, and, were they to stand alone, they would have precious little cumulative effect, amounting to little more than a pile of bricks awaiting an architect. To provide the architect, this paper can look to the work of Michel Foucault that provides 'analytical toolkit' to interrogate relationship between health and social care professions and user groups and consequences for older people, communities and informal carers. Foucault's analytical frameworks are relevant for applied gerontology as he problematises health and care professions as an instrument of state governmentality, an agent that reproduces dominant state discourses on 'old age'. The author has argued elsewhere that the power relations in health and care institutions produce user groups as 'customers' (Powell, 2006a, b; Powell, 2017). Simply put, in using this language of the customer, health and care policies attempt to replace the 'care' and 'health' subject position with that of 'consumer'. Despite personalisation, which has implications

J. Powell, *New Perspectives on Health and Social Care*, International Perspectives on Social Policy, Administration, and Practice,
https://doi.org/10.1007/978-3-031-25432-1_7

for families and communities, it is normalised through discourses and social practices such as surveillance, power and domination despite its rhetoric. At the same time, Foucault has identified that power is both repressive and a creative relation underpinned by 'resistance' practices. The process by which older people are made subjects are related to powerful 'managerial' actors in health and social care. This chapter provides an analytical framework critique to problematise the consolidation of managerial power primarily drawing from examples in the United Kingdom and United States. It fundamentally questions the assumption that managerialism in the post-welfare state has evolved independently, in its work and social practices. Indeed, in the past several years, social policy in the United States under President Trump and under PM Boris Johnson in the United Kingdom have focused on the management of old age with particular emphasis on the reform of health and social care. In the case of the United Kingdom, the process of reform was imposed by central government backed up by exaggerated demographic arguments in communities which has a long history of the present (Phillipson, 2013; Powell, 2017). It must also be stressed that the community care system was also in a less systemic and coherent structure; this also led to calls for it to be radically reformed (Phillipson, 2013). The personalisation reforms offered the promise of greater choice and autonomy for care workers, families, communities and service users through the intensification of marketisation into areas traditionally, directly controlled by central government such as community care. Coupled with this, service industries such as professional social work have had a fundamental redefinition of their profession as managerial systems (Lewis & Glennerster, 1996; Phillipson, 2013). Indeed, the nature and philosophy of such care provision for older people has been the central cornerstone in debates within the popular imagination in recent years. The continuing resonance of this question is linked to the perceived role of the professional social worker as a bulwark against an encroaching tide of 'dependency' (Hughes, 1995). A distinctive feature of the welfare state has been the systematic introduction of new managerialism throughout the public sector and introduction of new sectors such as the third sector. Thus, Powell (2017) points not only to the substantial increase in the number of people working in welfare organisations with the title 'manager' but also to the transformation of many formerly professionalised roles, into 'hybrids', where a significant aspect of the occupational identity is managerial. This is increasingly the case, for example, with head teachers, general practitioners and care managers. Care managerialism constitutes a move away from direct social care towards 'monitoring' and 'assessment' (Powell, 2017). The price to be paid, however, is that the relationship between helping professionals and older people has been reduced to one of surveillance. The surveillance role evokes a 'surface' of reality as constructed as 'depth', whereby professional methods of surveillance are presented as 'concern' models. This act of observation confers a uniformity that emphasises the 'protective' role of the professional rather than the substantive requirements of older people at the centre of surveillance (or 'professional monitoring'). In the United States and the United Kingdom in particular, the transition from a top-down social policy that managed dependent populations through the welfare state (pensions, unemployment insurance, and healthcare) to a post-welfare or neo-liberal state has gained momentum in recent years. Currently, surveillance occurs more bottom-up: central control has been replaced by local power, management systems are inspired by consumer models to welfare and there is a reliance on risk assessment. For example, at the local level in the United Kingdom, managerialism is seen to challenge 'dependency' by promoting social relationships of 'partnership' and 'trust' between professionals and older people underpinned by drawing in families and friends in communities to provide care. Such social relationships are supposed to have arisen because the mixed economy of welfare embodies a diversification of political agendas and multi public, private and third sectors: the control of financial resources, the improvement of services and promotion of choice, changing the role of local authorities and councils, a reduction in public sector provision focus on social relationships between professionals and service users and increasing family care. Indeed, if we recall in the United Kingdom, the Department of Health's (1991) policy rhetoric has highlighted the contradictions of the role of the British care manager: 'The guiding principle of implementation should be to achieve the stated objectives of the care plan with the minimum intervention necessary. It should, therefore, seek to minimize the number of service providers involved' (Hughes, 1995: 94). At the same time, British care managers should be 'tailoring services to individual needs and promoting wider choice' (Hughes, 1995: 94). Hence, the process of choice has become privatised and contingent on consumer sovereignty that has an impact on how resources are mobilised, especially in 2009 with a worldwide recession. Comparatively, the United Kingdom, the United States and Canada have similar policy shifts towards community care, but this has lacked sufficient integration between the funding, administrative and ideological frameworks that support acute and long-term care (Powell, 2017). In 2008 in the United States at the beginning of the 'credit crunch', with the Medicare D, this highlighted a shift of emphasis to a market-based, deregulated medical policy. Crucial aspects of the George W. Bush administration plan were characteristically delegated to insurance and pharmaceutical companies. Indeed, a preoccupation with the medical challenges presented by Medicare D, underpinned by privatised and insurance-driven health provisions, has resulted in what Foucault (1973) may have observed as an expansion of the medical gaze into wider areas of health and social policy. The medical gaze refers here to discourses, languages and ways of seeing that shape the understanding of aging into questions that centre on, and increase the power of, the care and health professions and delegitimise other possibilities. A consequence is that areas of policy that may at first seem tangential to the medical project come to be reflected in its particular distorting mirror. The impingement of the medical gaze can also be seen in the policy debates concerning disadvantaged groups over a shrinking public purse. The impact of medicalised notions of aging and its construction as a 'threat' to other sections of the population can also be seen in the proposed rationing of Medicare coverage in American welfare policy. Here, medical care has both come to colonise notions of old age and reinforce ageist social prejudices to the extent that 'infirmity' has come to stand for the process of aging itself and Medicare its potential policy saviour. Obamacare which attempted at universal coverage of medical access was stripped away by the Trump administration on the premise that individuals pay for their own care. Notwithstanding this, managerial power then is seen as a key professional role that is fundamental to transforming social relations within a consumer-led economy. Care management has become a space for the surveillance of older people. It is argued here that such a managerial mechanism is not of instigating choice, empowerment and social inclusion, but is a professional institution for individual regulation and collective control of user groups: older people. Hence, there are two dimensions that are particularly important to this chapter on theorising aging and professional power. First, there is the use of a Foucauldian perspective to locate the discontinuities in the relationship between health and social care professionals with older people. Second, there is the question of power itself and its relevance to the emergence of a discourse articulated by professional managers who assess, probe and inspect a distinct populational group: an aging population.

A Foucauldian Analytical Framework: Medicalisation and Managerialism

Michel Foucault's (1977a, b, c) landmark works have importance to the social analysis of old age. First, his analysis of punishment and discipline and medicine and madness has relevance to the images of older people. Foucault (1973) described how subjects of knowledge such as 'criminals' and 'mentally ill' are constructed through disciplinary techniques, for example, the notion of the expert 'gaze' (p. 29). Old age is constructed in a similar way. Second, Foucault (1977a, b, c, 1980a, b) made it possible to analyse both the official discourses embodied in health and social care policies and those operating and implementing within society: 'care managers' and 'older people': 'It was a matter of analysing, not behaviors or ideas, nor societies and their "ideologies," but the problematizations through which being offers itself to be, necessarily, thought—and the practices on the basis of which these problematizations are formed' (1980a, b: 11). The works of Foucault (1967, 1972, 1976, 1977a, b, c) have problematised issues of madness and illness, deviance and criminality and sexuality. Foucault has problematised the role of the medical 'expert' that seems 'empowering' but is a contingent sociohistorical construction. The relevance to old age is the recognition that social practices 'define a certain pattern of "normalization"' (Foucault, 1977a, b, c: 72). Such social practices are judged by experts such as medical personnel and care managers who problematise older people via a process of assessment for services. Care managers are pivotal to Foucault's (1977a, b, c) analysis of 'panoptic technology'; they normalise judgement on older people: the judges of normality are everywhere. We are in the society of the teacher-judge, the doctor-judge, the educator-judge and the 'social worker'-judge; it is on them that the universal reign of the normative is based; and each individual, wherever he may find himself, subjects to it his body, his gestures, his behaviour, his aptitudes and his achievements. The carceral network, in its compact or disseminated forms, with its systems of insertion, distribution, surveillance and observation, has been the greatest support, in modern society, of the normalising power (Foucault, 1977a, b, c: 304). For Foucault (1977a, b, c), normalising power involves the dimensions of physical and biological discourses and how these are inserted on the human body. The aging individual is located in a political field saturated with power relations that 'render it docile and productive and thus politically and economically useful' (Smart, 1985: 75). Hence, the care manager plays a role in power relations as she or he 'take(s) responsibility for ensuring that individuals needs are regularly reviewed, resources are effectively managed . . .' (Department of Health, 1989: 21). To examine this further, some of the conceptual tools emanating from Foucault illuminates their relevance for understanding the relationship between health and social care professionals and older people. Using Foucault's work focuses on two interrelated areas of Foucauldian analysis: medicalisation and aging and surveillance and crucially resistance. Jaber F. Gubrium (1998) has suggested that medicalisation is an important foundation for understanding how public and professional perceptions of older people are created via truth discourse. For Foucault (1980a, b), '"truth" is linked in a circular relation with systems of power which produce and sustain it' (p. 133). All strategies that attempt to control older people involve the production and social construction of 'true' knowledge. Historically and before the prevalence of managerial systems, biomedicine played a key role in articulating truths about the social condition of older people (Katz, 1996). The relevance of this to Foucault's work is the way in which the gaze of truth constructs people as both subjects and objects of power and knowledge. In *The Birth of the Clinic*, Foucault illustrates how such a gaze opened up 'domain of clear visibility' (Foucault, 1973: 105) for doctors for allowing them to construct an account of what was going on inside a patient and to connect signs and symptoms with particular diseases. The space in which the gaze operated moved from the patient's home to the 'hospital'. This became the site for intensive surveillance, as well as the attainment of knowledge, the object of which was the body of patients. The identities of 'elderly people' have been constructed through expert discourses of 'decay' and 'deterioration', and the medical gaze helps to intensify regulation over older people to normalise and provide assessment and treatment for such notions (Foucault, 1977a, b, c; Katz, 1996). Medical discourse, under the guise of science, was part of a disciplinary project oriented to create a model individual, conducting his life according to the precepts of health, and creating a medicalised society in order to bring conditions of life and conduct in line with requirements of health (Cousins & Hussain, 1984: 151). For medical professions, this legitimises the search within the individual for signs, for example, that she or he requires intense forms of surveillance and ultimately processes of medicalisation. This permeates an intervention into aging lives because practices of surveillance befit older people because of the medical discourse permeation of 'it's your age'. Surveillance of older people enables biomedicine to show concern for their health and acquire knowledge about their condition. Hence, it constructs older people as objects of power and knowledge: 'This form of power applies itself to immediate everyday life which categorizes the individual, marks him by his own individuality, attaches him to his own identity, imposes a law of truth on him which he must recognize and which others have to recognize in him. It is a form of power which makes individuals subjects' (Foucault, 1982: 212). The process of observation objectifies particular older people as 'diagnoses began to be made of normality and abnormality and of the appropriate procedures to achieve . . . to the norm' (Smart, 1985: 93). In this way, studying and examining the body and mind of older people is intrinsic to the development of power relationships between health professions and older people as users: 'The probing technique is at the centre of the procedures that constitute the individual as effect and object of power, as effect and object of knowledge. It is the examination which . . . assures the great disciplinary functions of distribution and classification' (Rabinow, 1984: 204). The probing technique, argues Foucault, combines panopticism and normalisation and 'establishes over individuals a visibility through which one differentiates them and judges them' (Foucault, 1984a, b, c: 184). Foucault (1977a, b, c) argued an individual is established as a 'case' and may be judged, measured and compared with others, in his or her very individuality. This individual may also have to be trained, classified, normalised and excluded. Foucault places great emphasis on the processes of assessment and surveillance. These processes are key elements in managerial power in 2009. Despite the surface of community care policy of idealising empowerment, the depth of community care is part of a disciplinary strategy which extended 'control over minutiae of the conditions of life and conduct' (Cousins & Hussain, 1984). Paradoxically, the care manager became the 'great advisor and expert' (Rabinow, 1984: 283) in assessing older people for care services. In general, disciplines such as gerontology took its place alongside medical power in correcting, disciplining and normalising biomedical language of 'decaying' older people. Nevertheless, the managerial gaze has come to rival the medical gaze. The power of the care manager as a gerontological expert has supplemented medical power. Managerialism and the Resistance of Older People Scientific dominance may have helped shape the medicalised construction of aging identities, though it was not economical enough in its reach. Science has been bound up with 'risk' (Beck, 1994) and what Giddens (1991) calls the process of 'reflexivity': This manifests because of the loss of faith in the exercise of scientific 'power/knowledge'. The focus on risk has led to a situation in which science has been slowly supplemented with financial discourses, and what we see, in relation to care provision, is an intensification towards care management models and consumerism. Hence, the pervasive move to a mixed economy of welfare has produced an extraordinarily powerful discourse and affects treatment of older people as 'consumers' that has come to accompany and supplement medical discourses of old age in communities and impact on families (Powell, 2017). Indeed, the mistrust of scientific 'power/knowledge' as manifested in biomedicine is mirrored by uncertainties against care models as a means of finding a legitimised place for older people. The language of choice to erode dependency has been colonised by both medical and care discourses. Indeed, the surveillance of older people can be seen as economically productive for central and local government especially relying on councils and informal carers in local communities. For example, social policy legislation in the United Kingdom, such as the recent integrated health and social care, still centres on a 'mixed economy of welfare' that highlights the incorporation of market forces to the construction and delivery of services (Clarke, 1994; Lewis & Glennerster, 1996; Phillipson, 1998). Hence, the mixed economy of welfare arguably fabricates representations of 'empowerment' for older people. For example, many people's needs have not been met due to power relations and agism (Bytheway, 1995). Allen, Hogg and Peace (1992) quoted a manager as saying: 'It is hard to listen to older people. They are slow in speech and thought' (p. 35). In this case, services provide schemes for the 'conduct of conduct' (Foucault, 1976) dominated by power/knowledge and characterised by the discretionary autonomy of managers of the state: 'It is within this disciplinary duality of power/knowledge and autonomy that power operates over older people, ultimately reinforcing the fragmentation that surveillance engenders in the broken identities of many older people at the centre of the professionals' gaze' (Powell, 2006a, b: 136). Indeed, power relationships are still constructed around barriers of marginalisation and dependency. As Henwood (1995) pointed out, pressures on resources were leading to reduced levels of service and a tightening of 'eligibility' criteria for older people. The American gerontologist Estes (1979) powerfully flags this up in her pioneering book *The Aging Enterprise*: 'Service strategies . . . and those for the aged . . . tend to stigmatise their clients as recipients in need, creating the impression that they somehow failed to assume responsibility for their lives. The needs of older persons are reconceptualised as deficiencies by the professionals charged with treating them' (p. 65). The point Estes is making is that such characteristics were illuminated as strategies for the service provision for older people in the United States. Terms such as 'frailty' are being used to define 'service eligibility', and power relationship processes are involved in service delivery and assessment (Estes, 1979). Hence, much discussion on the public health agenda in the United Kingdom in the past couple of years has overlooked the economic colonisation of frailty (Phillipson, 2013). Similarly, Foucault (1977a, b, c) sees the assessment as central a technique that renders an individual an object of power/knowledge. In the assessment leading the opening for social services, the statement of an 'aging body' is established in relation to normalised standards of rights and risks. Thus, older people will be probed for social, psychological and economic factors such as frailty, financial resources and expected levels of grades of supervision. This probe of assessment 'indicates the appearance of a new modality of power in which each individual receives as his status his own individuality, and in which he is linked by his status to the features, the measurements, the gaps, the "marks" that characterise him and makes him or her a "case"' (p. 192). For example, following Obamacare in the United States, certain aging identities are marked out for surveillance throughout the remainder of his or her chosen service by the Trump administration. Such a service can also be difficult to access now given it is a rich cohort who can tap those services with rich resources. There is an uncertainty about what sort of entitlements medical entitlement implies under Donald Trump as President, and how permanent those entitlements might be to older people as users. Two basic approaches to reform have arisen, and neither of them work. The first is to try to fix Obamacare (universal coverage) through corporate-style management. The second is to accept that Obamacare has failed the perceived test of affordability and to introduce a private tier of service delivery. This service delivery process impinges on balancing budgets through managerial assessments of older people who may or may not afford medical coverage. Heating or eating is a choice many have to choose, never mind medical care (Powell, 2017). The relevance of this to older people is that managerial power can intensify the ordering of identities through the processes of health and social care institutions and policies of the State. In a key research study still relevant today, Allen et al. (1992) found that most older people had only one or two services in their study. In this study, few older people had much choice in what services they received, any say in the time of the delivery, the person who delivered it or how much they received. Hence, this evidence highlights the numbing consequences are 'docile bodies' (Foucault, 1977a, b, c) drained of empowered energy, reinforced by the attitudes of managers to aging that it is just that, 'your age', which requires an inspecting 'gaze' and assessment of needs from 'expert' care managers to older people as 'consumers'. This relational positionality implies a top-down relation rather than partnership approach to tapping personalised services. Michel Foucault (1977a, b, c) has described how 'techniques of surveillance' that occur in the 'local centres of power/knowledge' (e.g. in the relationships between older people and care managers) have an individualizing effect: '. . . individualisation is "descending"; as power becomes more anonymous and more functional, those on whom it is exercised tend to be more strongly individualised . . . In a system of discipline . . . (older people become more individualised than younger people)' (p. 193). Techniques of surveillance are calculated, efficient and specific that 'inspection functions ceaselessly. The gaze is everywhere' (Foucault, 1977a, b, c: 195). Foucauldian ideas can identify related mechanisms of surveillance: panopticism and normalisation and the probe of assessment. These mechanisms train and organise individuals for their daily routines. Foucault (1977a, b, c) saw Jeremy Bentham's panopticon as the dominant example of disciplinary technology. Bodies of people can be made productive and observable. Foucault (1977a, b, c) remarked: 'Is it surprising that prisons resemble factories, schools, barracks, hospitals, which all resemble prisons?' (p. 44). In the context of care, assessment has a preoccupation with monitoring and surveillance, and this is crystallised in official policy discourse (Powell, 2006a, b). The perfect disciplinary apparatus, according to Foucault (1977a, b, c), 'would make it possible for a single gaze to see everything perfectly' (p. 173). Foucault approaches the mechanism of panopticism as both efficient, because surveillance was everywhere and constant, and effective, because it was 'discreet', functioning 'permanently and in silence' (p. 177). It also provides the scope for the supervision of those who were entrusted with the surveillance of others. The technique of panopticism was incorporated into health and social care relationships recently so that older people could be observed by professional surveillance (Phillipson, 2013). Social service provision for older people has elements of this kind of surveillance. Supervision is hierarchical in the sense that older people are accompanied by management discretionary power that embraces monitoring, assessing and calculating older people even in the given out of resources for personalisation, i.e. personal budgets for social care. Councils need to be kept informed of progress of clientele to communicate this at formal review meetings to establish resource allocation to service spending planning on personalisation. Surveillance however of older people does not stop at this point, as a network of reciprocal power relationships has been created: 'This network "holds" the whole together and traverses it in its entirety with effects of power that derive from one another: supervisors, perpetually supervised' (Foucault, 1977a, b, c: 176–177). Older people who require personalised services are the objects of scrutiny within society, but for such clients requiring further health financial services, the gaze reaches further; it evidences a strategic shift towards the surveillance of health and away from the post-war consensus relationship to respecting old age. Importantly, the question of resistance to professional power and its relationship to wider debates around autonomy is important in this chapter. Foucault (1976, 1977a, b, c) pointed to the different forms of opposition that have emerged in Western societies to challenge the imposition of power. The importance of consciousness and domination within care settings has formed the basis for group struggles for recognition and a badge of citizenship (Phillipson, 2013). As Clarke et al. (1978) point out, there exists dynamic relations between different groups: 'Negotiation, resistance, struggle; the relations between a subordinate and a dominant culture, whenever they fall within this spectrum are always intensely active . . . Their outcome is not given but made' (1977a, b, c: 44). The question of consciousness and resistance to professional domination has been tied to understanding and explaining social reality from older people through 'history from below'. In the United States, techniques of resistance to managerial power were found by Callaghan (1989) who claimed older people 'were particularly adamant that they did not want to be "cases" and no health and social care professionals needed to "manage" their lives' (p. 192). They wanted to be in control over their own lives.

Concluding Comments

The momentum for this Foucauldian analytical framework of health and social care and impact on care derives from the view that older people bear the indentation of principal modes of surveillance both in terms of medicalisation and managerial discourses and the decrease in resources to sanction in advanced capitalist societies and then the impactful emphasis on local communities, families and informal carers which creates new tensions of resources. Crucial queries have been raised as to the social relationships between professionals and older people and contract culture. To converse about social relationships in terms of 'contracts', for example, disguises the wider, often very hidden, power relationships that underpin and shape observable reality. Defining power relationships in discourses of consumers of services is problematic; it limits the power of the consumer and subtly alters the feasible grounds of complaint from collective concerns to the shortcomings of an individual transaction (the hallmark of neo-liberalism) and managerial power in the United Kingdom. Governments in the United States with President Donald Trump and the United Kingdom with PM Boris Johnson have focused on the management of old age with particular emphasis on the discourses of neo-liberalism into politically neutral and customer-safe questions with the service on offer from paying for medical coverage to UK councils and care managers with a focus on informal caring as the key to unlock saving the State resources. Following Foucault's (1977a, b, c) analysis of the relationship between power and knowledge, this change can be seen as the development of a matrix in which to speak seriously about professional power and older people the employment of discourse of surveillance would have to be entailed. The powerful language of surveillance Rethinking Community, Medicalization, Social and Health Care: A Foucauldian Analysis 34 Archives of Community and Family Medicine V2. I2. 2019 offers a form of universalism to health and care policy that has been subject to utter fragmentation. Rather than recognising patterns of social diversity, all are now equal under the monitor of care policy and practice. Contentiously, under such conditions, terms such as care manager and user have worn out their analytical usefulness except as a rhetorical disguise for those with legitimate power. They simply supply the masquerade whereby both health and social care professionals and older people would come to regulate themselves in the panopticists' theatre of surveillance. Further, families and friends in communities will also be entrusted with the care of older people as informal care saves the NHS billions of pounds every year in the United Kingdom. How do we know where we are going, until we know where we are coming from? A positioning of aging as requiring social care away from the State and a startling continuity of family and community care reinforced by medicalisation, surveillance by professional power which has gone un-challenged for a long time under the aegis of power should be critically reflected on without giving up the hope of resistance for active aging from the real experts society should be listening to and learning from the narratives from: older people themselves.

Chapter 8
Narrative, Health, Care and Family

Introduction

There has been an increasing interest in aging and family, within sociological developments relating to aging and social policy since the late 1990s to the present day. This is a trend that has cut across Canadian, American and European research (Cloke et al., 2006; Walker & Naegele, 1999; Bengtson et al., 2000; Biggs & Powell, 2001; Carmel et al., 2007). The reasons for such expansion are as much economic and political as they are academic. US and European governments recognise that the 'family' is important for social and economic needs, and this should be reflected in our understanding of aging, family processes and social policy. This leads to the question: how can we theoretically contextualise this and what are lessons for family research in sociological theorising? 'Narrativity' has become established in the social sciences, both as a method of undertaking and interpreting research (cf Kenyon et al., 1999; Holstein & Gubrium, 2000; Biggs et al., 2003) and as a technique for modifying the self (McAdams, 1993; Mcleod, 1997). Both Gubrium (1992) and Katz (1999) suggest that older people construct their own analytical models of personal identity based on lived experience and on narratives already existing in their everyday environments. By using a narrative approach, the meaning of family can be told through stories about the self as well as ones 'at large' in public discourse. 'Discourse' is an expression more often used to denote a relatively fixed set of stories that individuals or groups have to conform to in order to take up a recognised and legitimate role. Such an understanding of discourse can be found in the earlier work of Michel Foucault (1977a) and others. Self-storying draws attention to the ways in which family identities both are more open to negotiation and are more likely to be 'taken in' in the sense of being owned and worked on by individuals themselves. Families, of course, are made up of interpersonal relationships within and between generations that are subject to both the formal rhetoric of public discourse and the self-stories that bind them together in everyday life. The

J. Powell, *New Perspectives on Health and Social Care*, International Perspectives on Social Policy, Administration, and Practice,
https://doi.org/10.1007/978-3-031-25432-1_8

notion of family is, then, an amalgam of policy discourse and everyday negotiation and as such alerts us to the wider social implications of those relationships. Due to the mounting significance of current research in the areas of family and aging, the chapter investigates the particular relationship between old age and family life by means of studying the role of grand-parenting and the way it is perceived by the elderly, the family and the public at large. The study applies a narrative approach, hence telling the meaning of the family and grand-parenting through personal stories and public discourse, based on the theory of Michel Foucault. The findings put forth suggest that identities of family and grand-parenting are built on multiple grounds and that therefore theory should be sensitised accordingly, as identities are managed at different levels, for different audiences and at different levels of attentiveness and understanding. One finds here a source of raw material for the construction of identity and a series of spaces in which such identities can be legitimately performed. It is perhaps not overstating the case to say that the 'success' of a family policy can be judged from the degree to which people live within the stories or narratives of family created by it. In fact, the relationship between families and older people has been consecutively re-written in the social policy literature. Each time a different story has been told and different aspects of the relationship have been thrown into high relief. It might even be argued that the family has become a key site upon which expected norms of intergenerational relations and late-life citizenship are being built. This chapter explores the significance of such narratives, using developments in the United Kingdom as a case example that may also shed light on wider contemporary issues associated with old age. The structure of the chapter is fourfold. Firstly, we start by mapping out the emergence and consolidation of neo-liberal family policy and its relationship to emphasis on family obligation, state surveillance and active citizenship. Secondly, we highlight both the ideological continuities and discontinuities of the subsequent social democratic turn and their effects on older people and the family. Thirdly, research studies are drawn on to highlight how 'grand-parenting' has been recognised by governments in recent years, as a particular way of 'storying' the relationship between old age and family life. Finally, we explore ramifications for researching family policy and old age by pointing out that narratives of inclusion and exclusion often coexist. It is suggested that in the future, aging and family life will include the need to negotiate multiple policy narratives. At an interpersonal level, sophisticated narrative strategies would be required if a sense of familial continuity and solidarity is to be maintained. The Positioning of Neo-liberalism, Aging and Family Political and Social Debate since the Reagan/Thatcher years has been dominated by neo-liberalism, which postulates the existence of autonomous, assertive, rational individuals who must be protected and liberated from 'big government' and state interference. Indeed, Walker and Naegele (1999) claim a startling continuity across Europe is the way 'the family' has been positioned by governments as these ideas have spread beyond their original 'English-speaking' base. Neo-liberal policies on the family have almost always started from a position of laissez-faire, excepting when extreme behaviour threatens its members or wider social relations. Using the United Kingdom as a case example, it can be seen that that neo-liberal policy came to focus on two main issues. And, while both only represent the point at which a minimalist approach from the state touches family life, they come to mark the dominant narrative through which aging and family are made visible in the public domain (Cloke et al., 2006). On the one hand, increasing attention was paid to the role families took in the care of older people who were either mentally or physically infirm. A series of policy initiatives (UKG, 1981, 1989, 1990) recognised that families were a principal source of care and support. 'Informal' family care became a key building block of policy towards an aging population. It both increased the salience of traditional family values and independence from government and enabled a reduction in direct support from the state. On the other hand, helping professionals, following US experience (Pillemer & Wolf, 1986), became increasingly aware of the abuse that older people might suffer and the need to protect vulnerable adults from a variety of forms of abuse and neglect (Biggs et al., 1995). Policy guidance, 'No Longer Afraid: the safe-guard of older people in domestic settings', was issued in 1993, shortly after the move to seeing informal care as the mainstay of the welfare of older people. As the title suggests, this was also directed primarily at the family. It is perhaps a paradox that a policy based ostensibly on the premises of leaving-be combines two narrative streams that result in increased surveillance of the family. This paradox is based largely on these points being the only ones where policy 'saw' aging in families, rather than ignoring it. This is not to say that real issues of abuse and neglect fail to exist, even though UK politicians have often responded to them as if they were some form of natural disaster unrelated to the wider policy environment. To understand the linking of these narratives, it is important to examine trends tacit in the debate on family and aging, but central to wider public policy. Wider economic priorities, to 'roll back the state' and thereby release resources for individualism and free enterprise, had become translated into a family discourse about caring obligations and the need to enforce them. If families ceased to care, then the state would have to pick up the bill. It was not that families were spoken of as being naturally abusive. Neither was the 'discovery' of familial abuse linked to community care policy outside academic debate (Biggs, 1996). Discourses on the rise of abuse and on informal care remained separate in the formal policy domain. However, a subtle change of narrative tone had taken place. Families, rather than being seen as 'havens against a harsh world', were now easily perceived as potential sites of mistreatment, and the previously idealised role of the unpaid carer became that of a potential recalcitrant, attempting to avoid their family obligations. An attempt to protect a minority of abused elders thus took the shape of a tacit threat, hanging above the head of every aging family (Biggs & Powell, 2000). It is worth noting that these policy developments took little account of research evidence indicating that family solidarity and a willingness to care had decreased in neither the United Kingdom (Phillipson, 1998) nor the United States (Bengtson & Achenbaum, 1993). Further, it appeared that familial caring was actually moving away from relationships based on obligation and towards ones based on negotiation (Finch & Mason, 1993). Family commitment has, for example, to vary depending upon the characteristic care-giving patterns within particular families. Individualistic families provided less instrumental help and made use of welfare services, whereas a second, collectivist pattern offered greater personal support. While this study focused primarily on upward generational support, Silverstein and Bengtson (1997) observed that 'tight-knit' and 'detached' family styles were often common across generations. Unfortunately, policy developments have rarely taken differences in care-giving styles into account, preferring a general narrative of often idealised role relationships. It is not unfair to say that during the neo-liberal period, the dominant narrative of family became that of a site of care going wrong. Social Democracy, Aging and the Family Social democratic policies towards the family arose from the premise that by the early 1990s, the free-market policies of the Thatcher/Reagan years had seriously damaged the social fabric of the nation state and that its citizens needed to be encouraged to identify again with the national project. A turn to an alternative, sometimes called 'the third way', emerging under Clinton, Blair and Schroeder administrations in the United States and parts of Europe, attempted to find means of mending that social fabric, and as part of it, relations between older people and their families. The direction that the new policy narrative took is summarised in UK Prime Minister Blair's (1996) statement that 'the most meaningful stake anyone can have in society is the ability to earn a living and support a family'. Work, or failing that, work-like activities, plus an active contribution to family life began slowly to emerge, delineating new narratives within which to grow old (Hardill et al., 2007). Giddens (1998b) in the United Kingdom and Beck (1998) in Germany, both proponents of social democratic politics, have claimed that citizens are faced with the task of piloting themselves and their families through a changing world in which globalisation has transformed our relations with each other, now based on avoiding risk. According to Giddens (1998b), a new partnership is needed between government and civil society. Government support to the renewal of community through local initiative would give an increasing role to 'voluntary' organisations, encourages social entrepreneurship and significantly, supports the 'democratic' family characterised by 'equality, mutual respect, autonomy, decision-making through communication and freedom of violence'. It is argued that social policy should be less concerned with 'equality' and more with 'inclusion', with community participation reducing the moral and financial hazard of dependence (cf Walker, 2002; Biggs et al., 2003; Powell & Owen, 2007; Walker & Aspalter, 2008). Through an increased awareness of the notion of agism, the influence of European ideas about social inclusion and North American social communitarianism, families and older people found themselves transformed into active citizens who should be encouraged to participate in society, rather than be seen as a potential burden upon it (Biggs, 2001). A contemporary UK policy document, entitled 'Building a Better Britain for Older People' (DSS, 1998), is typical of a new genre of Western policy, re-storying the role of older adults on 'The contribution of older people is vital, both to families, and to voluntary organisations and charities. We believe their roles as mentors—providing ongoing support and advice to families, young people and other older people—should be recognised. Older people already show a considerable commitment to volunteering. The Government is working with voluntary groups and those representing older people to see how we can increase the quality and quantity of opportunities for older people who want to volunteer'. What is perhaps striking about this piece is that it is one of the few places where families are mentioned in an overview on older people, with the exception of a single mention of carers, many of whom, it is pointed out, 'are pensioners themselves'. In both cases, the identified role for older people constitutes a reversal of the narrative offered in preceding policy initiatives. The older person like other members of family structure is portrayed as an active member of the social milieu, offering care and support to others (Hardill et al., 2007). The dominant preoccupation of this policy initiative is not however concerned with families. Rather, there is a change of emphasis towards the notion of aging as an issue of lifestyle and as such draws on contemporary gerontological observations of the 'blurring' of age-based identities and the growth of the grey consumer (Katz, 1999). While such a narrative is attractive to pressure groups, voluntary agencies and, indeed, social gerontologists, there is, just as with the policies of the neo-liberals, an underlying economic motive which may or may not be to the long term advantage to older people and their families. Again, as policies develop, the force driving the story of elders as active citizens was to be found in policies of a fiscal nature. The most likely place to discover how the new story of aging fits the bigger picture is in government-wide policy. In this case, the document has been entitled 'Winning the Generation Game' (UKG, 2000a). This begins well with 'One of the most important tasks for twenty-first century Britain is to unlock the talents and potential of all its citizens. Everyone has a valuable contribution to make, throughout their lives'. However, the reasoning behind this statement becomes clearer when policy is explained in terms of a changing demographic profile: 'With present employment rates', it is argued, 'one million more over-50s would not be working by 2020 because of growth in the older population. There will be 2 million fewer working-age people under 50 and 2 million more over 50: a shift equivalent to nearly 10 percent of the total working population'. The solution, then, is to engage older people not only part of family life but also in work, volunteering or mentoring. Older workers become a reserve labour pool, filling the spaces left by falling numbers of younger workers. They thus contribute to the economy as producers as well as consumers and make fewer demands on pensions and other forms of support. Those older people who are not thereby socially included can engage in the work-like activity of volunteering. Most of these policy narratives only indirectly affect the aging family. Families only have a peripheral part to play in the story and do not appear to be central to the lives of older people. However, it is possible to detect the same logic at work when attention shifts from the public to the private sphere. Here the narrative stream develops the notion of 'grand-parenting' as a means of social inclusion. This trend can be found in the United Kingdom, in France, in Germany (Scharf & Wenger, 1995) as well as in the United States (Minkler, 1999). In the UK context, the most detailed reference to grand-parenting can be found in an otherwise rather peculiar place – namely, from the Home Office – an arm of British Government primarily concerned with law and order. In a document entitled 'Supporting Families' (2000b), 'family life', we are told, 'is the foundation on which our communities, our society and our country are built'. 'Business people, people from the community, students and grandparents' are encouraged to join a schools mentoring network. Further, 'the interests of grandparents, and the contribution they make, can be marginalised by service providers who, quite naturally, concentrate on dealing with parents. We want to change all this and encourage grandparents – and other relatives – to play a positive role in their families'. By which it is meant: 'home, school links or as a source of social and cultural history' and support when 'nuclear families are under stress'. Even older people who are not themselves grandparents can join projects 'in which volunteers act as "grandparents" to contribute their experience to a local family'. In the narratives of social democracy, the aging family is seen as a reservoir of potential social inclusion. Older people are portrayed as holding a key role in the stability of both the public sphere, through work and volunteering, and the private sphere, primarily through grandparental support and advice (Cloke et al., 2006). Grandparents, in particular, are storied as mentors and counsellors across the public and private spheres. However, the grandparental title has been used as a catchall within the dominant policy narrative, bringing with it associations of security, stability and in many ways an easier form of relationship than direct parenting; it exists as much in public as in private space. It is impossible to interpret this construction of grandparenthood without placing it in the broader project of social inclusion, itself a response to increased social fragmentation and economic competition. Indeed, it may not be an exaggeration to refer this construal of grandparenting as neofamilial. In other words, the grandparent has out-grown the family as part of a policy search to include older adults in wider society. The grandparent becomes a mentor to both parental and grandparental generations as advice is not restricted to schools and support in times of stress, but also through participation in the planning of amenities and public services (BGOP, 2000). This is a very different narrative of older people and their relationship to families, from that of the dependent and burdensome elder. In the land of policy conjuring, previously conceived problems of growing economic expense and social uselessness have been miraculously reversed. Older people are now positioned as the solution to problems of demographic change, rather than their cause. They are a source of guidance to ailing families, rather than their victims. Both narratives increase the social inclusion of a potentially marginal social group: formerly known as the elderly.

'Grand-Parenting' Policy and Care

There is much to be welcomed in this story of the active citizen elder. Especially so if policy-inspired discourse and lived self-narratives are taken to be one and the same. There are also certain problems, however, if the two are unzipped, particularly when the former is viewed through the lens of what we know about families from other sources. First, each of the roles identified in the policy domain, volunteering, mentor-ship and grand-parenting has a rather second-hand quality. By this is meant that each is supportive to another player who is central to the task at hand. Rather like within Erikson's psychosocial model of the lifecycle, the role allocated to older people approximates grand-generativity and thereby contingent upon the earlier, but core, life task of generativity itself (Kivnick, 1988). In other words, it is contingent upon an earlier part of life and the narratives woven around it and fails to distinguish an authentic element of the experience of aging. When the roles are examined in this light, a tacit secondary status begins to emerge. Volunteering becomes unpaid work; mentoring, support to helping professionals in their eroded pastoral capacities; and grand-parenting, in its familial guise, a sort of peripheral parent without the hassle. This peripherality may be in many ways desirable, so long as there is an alternative pole of authentic attraction that ties the older adult into the social milieux. Either that or the narrative should allow space for legitimised withdrawal from socially inclusive activities. Unfortunately, the dominant policy narrative has little to say on either count. Second, there is a shift of attention away from the most frail and oldest old, to a third age of active or positive aging, which, incidentally, may or may not take place in families. It is striking that a majority of policy documents of what might be called the 'new aging', start counting from age 50, an observation that is true for formal government rhetoric and pressure from agencies and initiatives lead by elders (Biggs, 2001). This interpretation of the life-course has been justified in terms of its potential for forming intergenerational alliances (BGOP, 2000) and fits well with the economic priority of drawing on older people as a reserve labour force (UKG, 2000b). Third, there is a striking absence of analysis of family relations at that age. Possibilities of intergenerational conflict as described in other literature (De Beauvoir, 1979), not least in research into three-generation family therapy (Hargrave & Anderson, 1992; Qualls, 1999), plus the everyday need for tact in negotiating childcare roles (Bornat et al., 1999; Waldrop et al., 1999), appear not to have been taken into account. This period in the aging life-course is often marked by midlife tension and multi-generational transitions, such as those experienced by late adolescent children and by an increasingly frail top generation (Ryff and Seltzer, 1996). Research has indicated that solidarity between family generations is not uniform and will involve a variety of types and degrees of intimacy and reciprocity (Silverstein & Bengtson, 1997). Finally, little consideration has been given to the potential conflict between the tacit hedonism of aging lifestyles based on consumption and those more socially inclusive roles of productive contribution, of which the 'new grand-parenting' has become an important part. While there are few figures on grandparental activity, it does, for example, appear that community volunteering among older people is embraced with much less enthusiasm than policymakers would wish (Boaz et al., 1999). Chambre (1993) claims volunteering in the United States diminishes in old age. Her findings indicate the highest rates of volunteering occur in mid-life, where nearly two thirds volunteer. This rate declines to 47% for persons aged between 65 and 74 and to 32 % among persons 75 and over. A UK Guardian-ICM poll of older adults indicated that, among grandfathers, but not grandmothers, there was a degree of suspicion of childcare to support their own children's family arrangements. More than a quarter of men expressed this concern, compared with only 19% of women interviewed. The UK charity, Age Concern, stated: 'One in ten grandparents are under the age of 56. They have 10 more years of work and are still leading full lives'. One might speculate, immersed in this narrative stream, that problematic family roles and relationships cease to exist for the work-returning, volunteering and community enhancing

50-plus 'elder'. Indeed, the major protagonists of social democracy seem blissfully unaware of several decades of research, particularly feminist research, demonstrating the mythical status of the 'happy family' (cf e.g. Land, 1999). What emerges from research literature on grand-parenting as it is included in people's everyday experience and narratives of self indicates two trends: (1) there appears to be a general acceptance of the positive value of relatively loose and undemanding exchange between first and third generations and (2) that deep commitments become active largely in situations of extreme family stress or breakdown of the middle generation. First, grandparents have potential to influence and develop children through the transmission of values. Subsequently, grandparents serve as arbiters of knowledge and transmit knowledge that is unique to their identity, life experience and history. In addition, grandparents can become mentors, performing the function of a generic life guide for younger children. This 'transmission' role is confirmed by Mills' (1999) study of mixed gender relations and by Waldrop et al.'s (1999) report on grandfathering. According to Roberto (1990), early research on grand-parenting in the United States has attempted to identify the roles played by grandparents within the family system and towards grandchildren. Indeed, much US work on grand-parenting has focused on how older adults view and structure their relationships with younger people. African American grandparents, for example, take a more active role, correcting the behaviour of grandchildren and acting like 'protectors' of the family. Accordingly, such behaviours are related to effects of divorce and under–/unemployment. Research by Kennedy (1990) indicates, however, that there is a cultural void when it comes to grand-parenting roles for many white families with few guidelines on how they should act as grandparents. Girard and Ogg report that grand-parenting is a rising political issue in French family policy. They note that most grandmothers welcome the new role they have in childcare of their grandchildren, but there is a threshold beyond which support interferes with their other commitments. Contact between older parents and their grandchildren is less frequent that with youngsters, with financial support becoming more prominent. Two reports, explicitly commissioned to inform UK policy (Hayden et al., 1999; Boaz et al., 1999), classify grand-parenting under the general rubric of intergenerational relationships. Research evidence is cited, that 'when thinking about the future, older people looked forward to their role as grandparents' and that grandparents looked after their grandchildren and provided them with 'love, support and a listening ear', providing childcare support to their busy children and were enthusiastic about these roles. Hayden et al. (1999) used focus groups and qualitative interviewing and report that 'grand-parenting included spending time with grandchildren both in active and sedentary hobbies and pursuits, with many participants commenting on the mental and physical stimulation they gained from sharing activities with the younger generation'. Coupled with this, the Beth Johnson Foundation found that older people as mentors had increased levels of participation with more friends and engendered more social activity. With the exception of the last study, each has relied on exclusive self-report data, or views on what grand-parenting might be like at some future point. In research from the tradition of examining social networks, and thus not overtly concerned with the centrality of grand-parenting or grandparent-like roles as such, it is rarely identified as a key relationship and could not be called a strong theme. Studies on the United Kingdom (Phillipson et al., 2000), Japan (Izuhara, 2000), the United States (Schreck, 2000; Minkler, 1999), Hispanic Americans (Freidenberg, 2000) and Germany, (Chamberlayne & King, 2000) provide little evidence that grandchildren, as distinct from adult children, are prominent members of older peoples reported social networks. Grandparental responsibility becomes more visible if the middle generation is for some reason absent. Thompson (1999) reports from the United Kingdom that when parents part or die, it is often grandparents who take up supporting, caring and mediating roles on behalf of their grandchildren. The degree of involvement was contingent however on the quality of emotional closeness and communication within the family group. Minkler (1999) has indicated that in the United States, 1 in 10 grandparents has primary responsibility for raising a grandchild at some point, with care often lasting for several years. This trend varies between ethnic groups, with 4.1% White, 6.55% Hispanic and 13.55% African American children living with their grandparents or other relatives. It is argued that a 44% increase in such responsibilities is connected to the devastating effects of wider social issues, including AIDS/HIV, drug abuse, parental homelessness and prison policy. Thomson and Minkler (2001) note that there is an increasing divergency in the meaning of grand-parenting between different socio-economic groups, with extensive caregivers (7% of the sampled population) having increasingly fewer characteristics in common with the 14.9% who did not provide childcare. In the United Kingdom, a similar split has been identified with 1% of British grandparents becoming extensive caregivers, against a background pattern of occasional or minimal direct care. It would appear that grand-parenting is not then a uniform phenomenon, and extensive grand-parenting or grandparent-like activities are rarely an integral part of social inclusion. Rather, while it is seen as providing some intergenerational benefit, it may be a phenomenon that requires an element of unintrusiveness and negotiation in its non-extensive form. When extensively relied on it is more likely to be a response to severely eroded inclusive environments and the self-protective reactions of families living with them. Minkler's analysis draws attention to race as a feature of social exclusion that is poorly handled by policy narratives afforded to the family and old age. There is a failure to recognise structural forms of inequality, and action seeking to socially include older people as a category appears to draw heavily on the occasional helper and social volunteer as a dominant narrative. Towards Diverse Narrative Streams? Each phase of social policy, be it the Reagan/Thatcherite neo-liberalism of the 1980s and early 1990s, the Clinton/Blairite interpretation of social democracy in the late 1990s or the millennial Bush administration, leaves a legacy. Moreover, policy development is uneven and subject to local emphasis and elision, which means that it is quite possible for different, even conflicting, narratives of family and later life to coexist in different parts of the policy system. Each period generates a discourse that can legitimate the lives of older people and family relations in particular ways and, as their influence accrues, create the potential of entering into multiple narrative streams. A striking feature of recent policy history has been that not only have the formal policies been quite different in their tenor and tacit objectives, one from another, but they have also addressed different areas of the lives of aging families. Where there is little narrative overlap, there is the possibility of both policies existing, however opposed they may be ideologically or in terms of practical outcome. Different narratives may colonise different parts of policy, drawing on bureaucratic inertia, political inattention and convenience to maintain their influence. They have a living presence, not least when they impinge on personal aging. Also, both policy discourses share a deep coherence, which may help to explain their coexistence. Each offers a partial view of aging and family life while downloading risk and responsibility onto aging families and aging identities. Neither recognises aging which is not secondary to an independent policy objective. Both mask the possibility of authentic tasks of aging. If the analysis outlined above is accepted, then it is possible to see contemporary social policy addressing diverse aspects of the family life of older people in differing and contradictory ways. Contradictory narratives for the aging family exist in a landscape that is a one and the same time increasingly blurred in terms of roles and relationships and split-off in terms of narrative coherence and consequences for identity. Indeed in a future of complex and multiple policy agendas, it would appear that a narrative of social inclusion through active aging can coexist with one emphasising carer obligation and surveillance. Such a coexistence may occasionally become inconvenient at the level of public rhetoric. However, at an experiential and ontological level, that is to say at the level of the daily lives of older adults and their families, the implications may become particularly acute. Multiple coexisting policy narratives may become a significant source of risk to identity maintenance within the aging family. One has to imagine a situation in which later lives are lived, skating on a surface of legitimising discourse. For everyday intents and purposes, this surface supplies the ground on which one can build an aging identity and relate to other family members and immediate community. However, there is always the possibility of slipping, of being subject to trauma or transition. Serious slippage will provoke being thrown onto a terrain that had previously been hidden, an alternative narrative of aging with entirely different premises, relationship expectations and possibilities for personal expression. Policy narratives, however, are also continually breaking down and fail to achieve hegemony as they encounter lived experience. Indeed, it could be argued that a continuous process of re-constitution takes place via the play of competing narratives. When we are addressing the issue of older people's identity in later life, we can usefully note Foucault's (1977a) contention that there has been a growth in attempts to control national populations through discourses of normality, but at the same time, this has entailed increasing possibilities for self-government. Part of the attractiveness of thinking in terms of narrative, that policies tell us stories that we don't have necessarily to believe, is the opening of a critical distance between description and intention. Policy narratives describe certain, often idealised, states of affairs. Depicting them as stories, rather than realities, allows the interrogation of the space between that description and experience (cf Powell, 2005).

Conclusion

What does this examination of social policy discourse and everyday stories of family and aging selves tell us, and what are the lessons for future sociological research? Firstly, we are alerted to the partial nature of the narratives supplied by social policy, which affects our perception of families as well as of older people. The simplifying role of policy discourse tends to highlight certain, politically valued, aspects of experience to the exclusion of other possibilities. These are also the discourses most likely to be reflected in policy sponsored research. Secondly, the inclusion of certain roles, activities and age bands in policy discourse has a legitimising role. In other words, it not only sanctions the direction of resources and the action of helping professionals important though that is. It also contributes to the legitimated identities afforded to people in later life. This includes at least two factors key to aging identity: the creation of social spaces in which to perform aging roles and be recognised as such and the supply of material with which explicit yet personal narratives of self and family can be made. Thirdly, a significant element in the 'riskiness' of building aging and family identities under contemporary conditions may arise from the existence of multiple policy discourses that personal narratives, of family, self and relations between the two, have to negotiate. Research on the management of identity should, then, be sensitised to the multiple grounds on which identity might be built and the potential sources of conflict and uncertainty may bring. Fourthly, attention should be paid to the relationship between tacit and explicit influences on identity management in late-life families. The multiple sources for building stories 'to live by' and the tension between legitimising discourses and alternative narratives of self and family would suggest that Narrative and Aging – The Social Construction of Grandparenting: A Foucauldian Analysis Open Journal of Geriatrics V2 . I2 . 2019 20 identities are managed at different levels, for different audiences and at different levels of awareness. There are implications here both for the conceptualisation of familial and policy relations and for the practice of research.

Chapter 9
Health and Care in the 'Risk Society'

Introduction

This chapter explores the relationship of health and social care to the concept of 'risk' that is both an epistemological tool and major facet of 'late modernity' (Delanty, 1999; Giddens, 1991). During the 1970s, the use of the notion of risk was mainly confined to 'natural sciences', where the concept was used to analyse and improve the 'security' of technological systems (Giddens, 1990). According to Lupton (1999), it was not until the 1980s and 1990s that social science-based disciplines discovered the importance of risk in relation to changes affecting modern society. In particular, the disciplinary development of sociology, for example, has discovered risk as one of the important aspects of neo-liberalism and modernity (Beck, 1992; Giddens, 1990; Luhmann, 1993; Delanty, 1999). The best-known approach in recent sociology of risk is the perspective of 'risk society' (Beck, 1992). This approach had a very large initial impact, but conceptual and empirical critiques have developed subsequently. Sociocultural research suggests the idea of a subject that is itself strongly influenced by its cultural context and builds up its own risk-knowledge referring to different, competing and sometimes contradictory knowledge systems which are available in different life situations and stages. For this reason, expert knowledge is only one point of reference among others (Delanty, 1999). People build up such 'private' knowledge on the base of their experiences during their life-course and in interaction with their contexts, others, the media, science and expert knowledge (Wilkinson, 2001).

Indeed, in contemporary Western society, risk is a broad concept that extends over a broad range of social practices that impinge on the experiences of older people. Current debates about older people and their relationship to sexuality, crime, national security, food safety, employment and welfare are all under-scored by risk (Phillipson, 1998). Moreover, there is an increasing recognition that potential risks of the present and future (global warming, genetic cloning, genetically modified

J. Powell, *New Perspectives on Health and Social Care*, International Perspectives on Social Policy, Administration, and Practice,
https://doi.org/10.1007/978-3-031-25432-1_9

foods and bio-terrorism) shatter the rigid boundaries of nations and demand global cooperation and control.

Awareness of the transboundary nature of risk has led the United Nations to form its own Commission on Human Security. A recent report by the UN Commission suggests ways in which the security of older people, for example, might be advanced, from humanitarian and military strategies through to economic, health and educational strategies. While 'freedom from want' continues to be the most pressing global imperative, in recent years, 'freedom from fear' has risen up the global political agenda (Commission on Human Security, 2003: 4, cited in Powell & Wahidin, 2004). Coupled with this, the US Central Intelligence Agency's 2004 World Factbook suggests that a 'health and social care population' is a risk to the financial safety of Western nation states (Powell & Wahidin, 2004).

In science, risk has traditionally been approached as an objective material entity, to be mastered by processes of calculation, assessment and probability. In the twenty-first century, 'advances' in science and medicine led to the eradication of many infectious diseases, raised life expectancy in old age and improved quality of life. Accordingly, it is no surprise that the literature on risk was housed within the confines of specialist expert disciplines, such as biomedical gerontology (Timiras, 1997).

The nature of scientific knowledge about risk and its impact on health and social care has articulated the perspective that, as a person goes through the health and social care process, there are heightened risks to the human body, in the mind and in the internal organs (Hughes, 1995). It has gradually become clear that the very institutions entrusted with regulating risks have themselves transmuted into risk producers. In recent times, multinational corporate business, science, medicine and government have all been accused of generating various dangers to public health that threaten the safety of older people. The rising cultural profile of risk has brought to the fore deep-rooted ethical concerns about the relationship between individuals, institutions and society (Lupton, 1999: 112). The debate currently taking place about the use of reproductive technologies for human cloning to prolong life in old age stands as a case in point. Such situations indicate that, in addition to being a scientific and economic issue, risk is now also interpreted as both social and political issues.

It has become commonplace for academics and practitioners to explore, develop and apply an assortment of social science perspectives on risk. In a post-9/11 world, questions around risk and risk management have become ever more pertinent, leading to reflections on a number of different levels about 'ontological security'. Moreover, how do older people manage their sense of well-being in a world in which less and less can be taken for granted? To what extent does the spectre of global risks interplay with more mundane insecurities that reach to the capillary texture of the day-to-day life of older people?

In a climate of indeterminacy, there is an urgent need to draw out the ways in which wider social theory can elucidate the dynamic relationship between risk and health and social care in contemporary society. In response to public concerns about un-bounded technoscientific development and the apparent inefficacy of expert systems, interest in risk has gathered momentum within social science disciplines in recent years (cf. Giddens, 1991). However, while the language of risk has become prolific, the concept itself remains cloaked in ambiguity and its relationship to health and social care scantily researched, making risk and health and social care an important and significant issue for social theory and social gerontology.

A theoretically informed understanding of risk illustrates the interconnectedness of a 'health and social care population', social policy and social life. From this perspective, risk is more than an estimation of costs and benefits; it is a theoretical instrument for weighing dissimilar sets of values and political orientations that affect the positioning of individuals and populations. As Niklas Luhmann (1993: ix) points out: 'How do we comprehend our society if we turn the concept of risk into a universal problem neither avoidable or evadable? What is now necessary? And accordingly: What is chance? How does society in the normal performance of its operations cope with a future about which nothing certain can be discerned, but only what is more or less probable or improbable?'

The deployment of risk to gerontological contexts has facilitated disruptions of both the metanarratives inscribed within gerontology, as both discipline and policy, and the ontological status of its subjects, that is, older people. The rationale for this argument is not about a normative foundation for articulating agency afforded to older people. Part of an understanding of individual freedom is how human action is plugged into and (de)regulated by a social context (Mills, 1959). Such an approach seeks to capture the dimensions of subjectivity within the sociopolitical constraints that shape individual lives. This allows reconstructions of logics of action or structuration behind current neo-liberal self-representations of health and social care identity. It could be supposed that such constructions enable us to reconstruct the complexity of health and social care in social contexts and the influence of, for example, health and social care on these experiences as a ground for risk perception. Hence, we need to understand the major social forces that impinge on health and social care itself. Such social forces that create risk associated with health and social care imply a breakdown in trust as a key modernist principle in contemporary society.

From Trust to Risk

There are increasing attempts to conceptualise the notion of 'trust' in social theory as a pivotal dimension of modernity (Giddens, 1991). Trust is incompatible, on the one hand, with complete ignorance of the possibility and probability of future events, and, on the other hand, with emphatic belief, when the anticipation of disappointment is excluded. Someone who trusts has an expectation directed to an event. The expectations are based on the ground of incomplete knowledge about the probability and incomplete control about the occurrence of the event. Trust is of relevance for action and has consequences for the trusting agent if trust is confirmed or disappointed. Thus, trust is connected with risk (Giddens, 1991).

Up to now, there have been few attempts to work out a systematic scheme of different forms of trust between older people and individuals, institutions or policies that have bearing on their identity performance. Social trust tends to be high among older people who believe that their public safety is high (Wahidin & Powell, 2001). As public trust erodes regarding institutions like the government and the media (Phillipson, 1998), trust attracts more and more attention in social sciences.

Mölling (2001) distinguishes between trust in contracts between people and state, such as pension provision, trust in friendships across intergenerational lines, trust in love and relationships and trust in foreign issues associated with national identity. However, sociological theories that suppose a general change in modernity (cf. Beck, 1992) assume that, with the erosion of traditional institutions and scientific knowledge, trust becomes an issue more often produced actively by individuals than institutionally guaranteed.

Independent from the insight that social action in general is dependent more or less on trust, empirical results in the context of risk perception and risk taking indicate the following:

- Trust is much easier to destroy than to build.
- Once trust is undermined, it is more difficult to restore it.
- Familiarity with a place, a situation or a person produces trust.
- People will develop trust in other people or situations if those people or situations have positively valued characteristics.

Trust seems to be something that is produced individually by experience and over time and cannot be immediately and with purpose be produced by organisations or governments without dialogical interaction with older people on issues affecting their lifestyles and life-chances such as care, pensions, employment and political representation (Walker & Naegele, 1999). However, as Giddens (1991) stresses, risk is a feature of a society shifting its emphasis away from trust in traditional ties and social values. How risks are perceived and formulated as a breakdown in trust reflects the essentially discursive practices of politics and power in modern society. The ability to control and manage perceptions about moral intentions of a pervasive governmental rationality may be part of an understanding of risk.

For this reason, risk is an evocative and compelling focus of theoretical analysis, requiring for understanding a historical backdrop for the emergence of a risk society in the West that influences the identities of older people.

The Historical Rise and Consolidation of 'Risk'

Over history, the word risk has changed its meaning and has become far more common applied to the global and the local (Beck, 1992). Most commentators link the emergence of the word and concept of risk with early maritime ventures in the premodern period. Ewald (1993) argues that the notion of risk first appeared in the Middle Ages, related to maritime insurance, and alluded to the perils that could compromise a voyage: 'At that time, risk designated the possibility of an objective danger, an act of God, a force, a tempest or other peril of the sea that could not be imputed to wrongful conduct' (Ewald, 1993: 226).

Luhmann (1993: 9) claims that the German word for risk appeared in the mid-sixteenth century and the English, in the second half of the seventeenth. He notes that the Latin term riscum had been in use long before in Germany and elsewhere.

Importantly, this premodern concept of risk excluded the idea of human fault and responsibility. Risk was perceived to be a natural event, such as a storm, flood or epidemic, rather than a human-made one. By the eighteenth century, the concept of risk had begun to be rationalised under the new scientific paradigm, drawing upon new ideas of mathematical probability. The development of statistical calculations of risk and the expansion of the insurance industry in the early modern era meant that 'Consequences that at first affect only the individual became "risks," systematically caused, statistically describable and in that sense "predictable" types of events, which can therefore also be sub-jected to supra individual and political rules of recognition, compensation and avoidance' (Beck, 1992: 99).

By the late nineteenth century, the notion of risk was extended. It was no longer located exclusively in nature but was 'also in human beings, in their con-duct, in their liberty, in the relations between them, in the fact of their association, in society' (Ewald, 1993: 226). The development of this concept assumes human responsibility and that 'something can be done' to prevent misfortune. Feelings of insecurity are common, just as they were in premodern times, but we now harbour somewhat different fears, and different objects are causes for our anxiety.

As in premodern times, we are aware of certain types of risks and possible outcomes, but our responses are more governed by a scientific rationale. The symbolic basis of our uncertainties is fundamentally created by disorder or insecurities, such as the loss of control over our bodies, our relationship with others and self and the extent to which we can exert autonomy in our everyday lives. This body of governance or strategies can take many forms, from dieting to fitness regimes to minimising potential risk to self by taking out life insurance, installing a burglar alarm, etc. These are just some of the ways by which the heightened awareness of everyday risk is negotiated, contained and managed. Therefore, rational thinking, systems of prevention and ways of identifying threats before they take effect are means of health and social care danger and threats in a risk society. These strategies are the products of late modern ways of thinking about, and reacting to, risk.

Every technology produces, provokes and programmes a specific accident. The invention of the boat was the invention of shipwrecks. The invention of the steam engine and the locomotive was the invention of derailments. The invention of the highway was the invention of 300 cars colliding in 5 min. The invention of the airplane was the invention of the plane crash. I believe that from now on, if we wish to continue with technology (and I don't think there will be a Neolithic regression), we must think about both the substance and the accident (Virilio, 1983: 32).

The modernist concept of risk represented a new way of viewing the world and its chaotic manifestations and its contingencies and uncertainties. It assumed that unanticipated outcomes may be the consequence of human action rather than 'expressing the hidden meanings of nature or ineffable intentions of the Deity', largely replacing earlier concepts of fate or fortune (Giddens, 1990: 30). As Reddy argues: 'Moderns had eliminated genuine indeterminacy, or "uncertainty," by inventing "risk." They had learnt to transform a radically indeterminate cosmos into a manageable one, through the myth of calculability' (Reddy, 1996: 237). Castel goes even further, arguing that the obsession with the prevention of risk in modernity is built upon a grandiose technocratic rationalising dream of absolute control of the accidental, understood as the irruption of the unpredictable. A vast hygienist utopia plays on the alternate registers of fear and security, inducing a delirium of rationality, an absolute reign of calculative reason and a no less absolute prerogative of its agents, planners and technocrats, administrators of happiness for a life to which nothing happens (Castel, 1991: 289).

In modernity, risk, in its purely technical meaning, came to rely upon conditions in which the probability estimates of an event are able to be known or knowable. The use of risk in common discourse litters texts, images and day-to-day conversation. In contemporary Western societies, the noun risk and the adjective risk have become very commonly used in both popular and expert discourses. Risk and uncertainty tend to be treated as conceptually the same thing. The term tends to be used to refer almost exclusively to a threat, hazard, danger or harm: we 'risk our life savings' by investing on the stock exchange, or we 'put our health at risk'. In this sense, risk means somewhat less than a possible danger or a threat and more an unfortunate or annoying event. Risk is therefore a very loose term and, as the above shows, is used in a variety of ways. Issues of calculable probability and economic systems of accountability are not necessarily important to the colloquial use of risk. Risk knowledges are historical and local, in the literature known as glocal. What might be perceived to be 'risky' in one era at a certain locale may no longer be so viewed later, or in a different place (Phillipson & Powell, 2004). As a result, risk knowledges are constantly contested, subject to disputes and debates over their nature, their control and who is to blame for their creation.

Over time the constellations of risks have changed. The risks of early industrialisation were evident: they could be smelt, touched, tasted or observed with the naked eye. These hazards 'assaulted the nose or the eyes and were thus perceptible to the senses' (Beck, 1992: 21). Earlier risks were part of a system of stratification and poverty that was highly visible; in contrast, today's hazards are invisibly everywhere in the everyday. Many of the major risks today largely escape perception 'for they are localized in the sphere of physical and chemical formulas (e.g., toxins in foodstuffs, toxins in the environment, [certain pollutants] or nuclear threat)' (Beck, 1992: 21). These risks exist in scientific knowledge rather than in everyday experience. Expert knowledges tend to contradict each other, resulting in debates over standpoints, calculation, assessment procedures and results. This has the effect of paralyzing action and bringing insurance systems that promised to cover eventualities into chaos. In Great Britain, for example, the welfare state, an insurance system that promised to care for people from cradle to the grave, is unable to sustain that promise for future generations. The welfare system as a system of social insurance is beginning to lose its legitimacy with the rise of private health insurance.

Scientists have lost their authority in relation to risk assessments, most evidently seen in the collapse of endowment and certain pension funds. Scientific calculations are challenged more and more by political groups and activist (Beck, 1994: 125–26). The nature of such hazards, therefore, returns the concept of risk to the premodern notion of 'incalculable insecurities'. In common with such hazards, insecurities 'undercut the social logic of risk calculation and pro-vision' (Beck, 1994: 77). For Beck, then, risk may be defined as a 'systematic way of dealing with hazards and insecurities induced and introduced by mod-ernisation itself' (Beck, 1992: 21). Dangers and hazards in modern society are the cause and effect of humanly generated productions and in essence can be avoided, monitored or altered. Therefore, the above debates demonstrate that the essence of risk is not that it is happening, but that it might be happening.

These debates include 'reflexive modernisation' (cf. Giddens, 1991), or the move toward criticism of the outcomes of neo-liberalism, and individualisation, or the breaking down of traditional norms and values. Older people living in neo-liberal societies have therefore moved towards a greater awareness of risk and are forced to deal with risks on an everyday basis: 'Everyone is caught up in defensive battles of various types anticipating the hostile substances in one's manner of living and eating' (Beck, 1992: 45). The media, for their part, have taken up warnings of experts about risk and communicate them to their mass publics. They have also reported disputes among these experts that concern risks: how serious they are, who should be blamed for them and what the most appropriate course of action might be. In risk society, therefore, the politics and sub-politics of risk definition become extremely important. They highlight the contested nature of who is defining what is a risk and how. Various specialised fields have developed around 'risk'. Our understanding of risk derives from pervasive references to risk behaviour, risk awareness, risk analysis, risk assessment, risk communication and risk management at any given time. Such terms allude to major fields of research and practice, developed in an attempt to measure, evaluate and regulate risk in areas as far-ranging as medicine, law, education, social policy and health, but they also signify the techniques by which, as social actors, we come to make sense of the everyday. In the last few years, there has been a series of 'panics' about welfare resources that has revolved around particular items in the United Kingdom, such as the effects of a 'health and social care population' on public resources and taxation. Thus, perceptions of risk are intimately tied to understandings of what constitutes a danger and for whom. This im/materiality or the invisibility gives risk an air of unreality until the moment at which they materialise as symptoms or as irreversible side effects.

The growing disparities between rich countries and poorer ones lead to growing inequalities in risk distribution, where class positions and risk positions overlap. As we see these disparities develop in the new global order, some risks, such as those caused by hazardous industries, are transferred away from the developed countries to the Third World. Thus, while Beck sees risk society as a catastrophic society, what we are seeing is the transference of certain risks through aversion and management, which in turn include a reorganisation of power and authority (Beck, 1992: 4).

The 'Risk Society' Thesis

Beck (1992) in his work *Risk Society* acknowledges that some social groups are more affected than others by the distribution and growth of risks, and these differences may be structured through inequalities of class and social position. The disadvantaged have fewer opportunities to avoid risk because of their comparative lack of resources. By contrast, the wealthy to a degree (income, power or education) can purchase safety and freedom from risk (Beck, 1992: 33). However, it is the gestation and the constellations of the risks that are unknown, and thus risk also affects those who have produced or profited from them, breaking down the previous social and geographic boundaries evident in modern societies.

Beck argues that the 'former colonies' of the Western nations are soon becoming the waste dumps of the world for toxic and nuclear wastes produced by more privileged countries. Risks have become more and more difficult to calculate and control. Hence, it can be argued that risks often affect both the wealthy and poor alike: 'poverty is hierarchic and smog is democratic' (1992: 36). At the same time, because of the degree of interdependence of the highly specialised agents of modernisation in business, agriculture, the law and politics, there is no single agent responsible for any risk: '[T]here is a general complicity, and the complicity is matched by a general lack of responsibility. Everyone is cause and effect' (33) and so 'perpetrator and victim become identical' (38) in a consuming society. It is this im/materiality and in/visibility of the threats that saturate the 'risk society', making it harder to identify the offender in global risk. Lupton (1999) argues that this fundamentally poses the second challenge for analyses of these socially constituted industrial phenomena: interpretation becomes inherently a matter of perspective and hence political. Politicians constantly invoke science in their attempts to persuade the public that their policies and products are safe. The inescapability of interpretation makes risks infinitely malleable and, as Beck (1992: 23) insists, 'open to social definition and construction'. This in turn put those in a position to define (and/or legitimate) risks, the mass media, scientists, politicians and the legal profession, in key social positions.

'Risk society', argues Beck, 'is not an option which could be chosen or rejected in the course of political debate' (1996: 28, Authors italics). Instead, this is an inescapable product and structural condition of advance industrialisation in which the produced hazards of that system, in Beck's words (1992: 31), 'undermine and/or cancel the established safety systems of the provident state's existing risk calculation'. Beck (1992) in his work further emphasises this point by examining contemporary hazards associated with nuclear power, chemical pollution and genetic engineering and biotechnology that cannot be limited or contained to particular spaces, and that which cannot be grasped through the rules of causality, and cannot be safeguarded, compensated for or insured against. They are therefore 'glocal': both local and global. Risk society is thus 'world risk society', and risks affect a global citizenship. The questioning of the outcomes of modernity in terms of the production of risks is an outcome of reflexive modernisation. Our awareness of risk, therefore, is heightened at the level of the everyday.

Both Beck (1992) and Giddens (1991) have a commonality of exploring and assessing risk linked to personal conduct and ontological security. Both claim that modernisation helps the self become an agent via processes of individualisation which they both see as indicative of neo-liberalism; they advocate that the self become less constrained by structures and become a project to be reflexively worked on.

Further studies refer to general discourses and their influence of problem definitions and the constitution of groups 'at risk' as older people (Estes et al., 2003). Such studies show how generalised social categories in institutional and media discourses produce homogenous groups in relation to risk. On the one hand, these categories do not do justice to the diverse persons behind the categories, and a new reality is entailed that changes the whole notion of a social group. For example, risk anxiety engendered by the desire to keep older people safe may restrict their autonomy and their opportunities to develop necessary skills to cope with the world.

Notwithstanding this, there is a criticism that 'risk' is narrowed to the responses of technical and environmental risks as unforeseen consequences of industrialisation. This concept of a danger-consequence society (Massumi, 1993) fails to grasp the more general societal development regarding the concept of risk as a specific historical strategy to manage uncertainties. This strategy is strongly linked to the idea of insurance and the statistical methods to calculate uncertainties developed in modernity (Hacking, 1990). Many risk theorists share this view (Giddens, 1991) but support a more general notion on risk and risk responses in current societies concerning the ways in which uncertainties of health and social care populations are managed in general. The narrowed view on technical and statistical risk management seems to be insufficient for the given complexity concerning, for example, governmental risk strategies and rationalities (Rose, 1996) and emotional spaces. Further critique aims at the assumption that new risks produce a general anxiety that would support a higher public awareness of risk and involve increasing the political commitment of the public. It has been argued that this does not apply to all risks, and neither do all people respond in the same way (Lupton, 1999).

The existing literature on risk ignores the role of emotion in theorising. Often it is assumed without further empirical examination that there is a direct link between social structure and emotion or risk consciousness and societal anxiety of 'emotions'. One clear example of risk associated with emotions and health and social care is crime and victimisation. As with other characteristics that make older people vulnerable to victimisation, it is difficult to disentangle the age factor from other variables that mean that older people figure prominently among those for whom victimisation has a high risk. In terms of actual rates of victimisation, elder abuse is certainly underreported (Biggs, 1993). Many older people may feel vulnerable related to crime and abuse (Wahidin & Powell, 2001). Older victims tend to report that crime has a high and long-lasting impact upon them compared to younger victims.

Fattah and Sacco (1989) conclude in their study of crime against older people in North America that 'While it may be fashionable to view fear of crime as an irrational response on the part of elderly to a world that does not truly threaten them, such a conceptualisation is probably not appropriate. Rather than irrationality, elderly fear of crime may represent the exercise of caution by a group in society that frequently lacks the control necessary to manage the risk of criminal harm or to marshal the resources necessary to offset its consequences' (1989: 226).

Hence, the multilayered results of the link between risk, crime and emotion show that additional theoretical work in social gerontology is necessary. Wahidin and Powell (2001) refer to the idea of a fundamental change in modernity: that there is also a reinterpretation of the uncontrollable and unforeseeable in something observable such as crime.

Health and Social Care in Risk Society: A Case Study Based on Health and Social Care

Throughout risk literature, there has been a growing gap in the application of risk in health and social care. Notwithstanding this, in order to work on itself, the 'self' relates to risk: 'risks become the motor of self-politicisation of modernity in industrial society'. How the conjunction between risk and individual awareness has become politicised provides a perspective on the viability of neo-liberalism and conditions of human action. One element of the 'motor' of self-politicisation is how successful neo-liberalism has been in fashioning common-sense discourses around its political rhetoric. Jürgen Habermas (1992) claims that what we are witnessing is a 'completely altered relationship between autonomous and self-organized public spheres on the one hand, and sub-systems steered by money and administrative power on the other'. Self-autonomisation coupled with administrative power is indicative of 'risk': neo-liberal features of social policy for older people.

There is then an ambivalence at the heart of neo-liberalism: on the one hand, older people are to be 'managed' by other administrative powers such as professional experts in modernity (Leonard, 1997); on the other hand, older people are left to govern themselves, a process that Rose and Miller (1992) call 'action at a distance' from the state. Hence, as consumers, older people are distanced further away from the state; rather than a cause for celebration, the dystopian implications are far-reaching in that they generate further risks that the self must negotiate with the withdrawal of the state.

Indeed, constituting risk as a centrally defining motif of 'late modernity' offers a new perspective: it allows the interrogation of how older people are made subjects. This inevitably impinges on risk. Neo-liberalism gives the impression that older people have the capacity to generate their own 'human agency' as indicative of 'consumer culture' (Gilleard & Higgs, 2000) irrespective of structural constraints. The problematic of self-governance, or what Nikolas Rose (1993) calls 'the government of freedom', needs critical reflection. Such reflection is sensitised to mapping out key features of neo-liberal political rationalities that impinge on 'self-government' within the 'risk society'. Notably those arising from the shift from welfarism to neo-liberalism.

We can draw from neo-liberal features of social policy to highlight how old age was socially and politically positioned as an opportunity for processes of government related to self-governance (Rose, 1993, 1996). Theoretical normativity lies at the heart of the 'structure vs. agency' debate as applied to social gerontology. Social gerontology as discipline and practice can be sensitised to how older people ought to have autonomy and control in their own lives; but this masks critical questions of the what is. As Powell (1998a, b) asks, how do we know where we are going until we know where we are coming from? So, we point out features of the 'risk society' and how this impacts on any historical, contemporary and future questioning of health and social care. The next section traces a genealogy of health and social care, risk and its interrelationship with social welfarism.

Genealogy of Risk and Health and Social Care Identities

There has been a change in the structure that has underpinned definitions of health and social care. The key development here concerned the way in which, in industrial societies, growing old was altered by the social and economic institutions connected with the welfare state. These became crucial in shaping the dominant discourse around which health and social care was framed. A key theme was the reordering of the life-course into life zones associated with education, work and retirement (Phillipson, 1982). A final element concerned the role played by services for older people as a measure of the move to a more civilised society. 'Old age' was itself the creation of modernity, reflecting the achievements of industrialism, improved public health and the growth of social welfare (Biggs, 1999). The steady growth in the proportion of older people in the population was, until the beginning of the 1980s, largely contained within the institution of the welfare state.

Indeed, older people have had the most to lose, given the restructuring of relationships associated with the emergence of a postmodern society. As Biggs (1993) argues, modern life raises at least two possibilities: the promise of a multiplicity of identities on the one side and the danger of psychological disintegration on the other. This development has served to change the definition of what it means to be an older person. In the conditions of advanced modernity, growing old moves from being a collective to an individual experience and responsibility. This new development may be seen as a characteristic of a society in which the 'social production of risk' runs alongside that associated with the 'social production of wealth' (Beck, 1992). Beck (1992: 21) further defines the nature of risk as a 'systematic way of dealing with hazards and insecurities induced and introduced by modernization itself' [emphasis supplied]. Such modernisation is shaped by neo-liberalism which in turn positions the role and responsibility of what it means to be a consumer in Western society.

Conclusion

The neo-liberal surface dominance in social policy has been very successful because it has identified existential concepts such as self-responsibility, self-governance and self-care which are to be used to facilitate human action and self-government (Powell, 2001a). The regulation of personal conduct is no longer the responsibility of the state.

Here neo-liberal discourse attempts to define the social policy domain to interpret valid human needs and limit rights, indicative of risk. Coupled with this, processes and relationships in the management of old age are decided by governmental rationalities that are tied to questions of self-governance, self-actualization and autonomy. Further, neo-liberalism considers that a welfare society must reflect only the interplay of social and political structures forged out of self-responsibility and consumerism (Leonard, 1997; Powell, 2001a). As we discussed in a previous chapter regarding analysis of Foucault's (1978a, b, c, d) notion of 'governmentality', older people as autonomous consumers derive their 'care', individually and collectively, from a range of social policies, institutions and sites, so that the organization of care involves market forces, families and state and care institutions. In the United States, responsibility for the administration of care-based social policy has been the contemporary case management institution (Powell, 2001a).

Case management as an institutional apparatus has been presented as consolidating neo-liberalism by adding 'choice' and reducing 'risks' and 'problems' associated with health and social care. This movement away from 'helping relationship' to 'care management' is an aspect of governance: that is, the mechanism by which clients, as individuals and collectively, are 'disciplined' away from 'structured dependency' to the governance of 'self' in the 'risk society' (Powell, 2001a). Such reforms were about recasting older people as consumers in a marketplace to be managed by administrative powers (Powell, 2001a; cf. Habermas, 1992).

Indeed, Biggs and Powell (2001a, b, c: 110) raise critical questions about the relationship between old age, neo-liberalism and new sources of oppression associated with the risk society: those who do not conform to the utopian dream appear to have been shunted into a non-participative discourse, bounded by professional surveillance or the more palatable yet closely related discourse of 'monitoring'. In both cases, it could be suggested that a discourse on dependency has been supplemented, and in some cases replaced, by a discourse on risk. The risk of giving in to a health and social care body, the risk of thereby being excluded from one's retirement community, the risk of being too poor to maintain a consumer lifestyle, the risk of being excluded from participation through incapacity that has been externally assessed, the risk of being abused, the risk of control being taken out of one's hands and the risk of tokenism in partnership.

Risk is then much more than a calculation of costs and benefits, it is a theoretic endeavour whereupon what seems to be a simple question such as 'Do we live in a risk society?' is one to which we are incapable of providing a straight answer. As consumers, has our duty of care become onerous or riskier as we journey into the stages of later life? Can anyone escape the miasma of a risk? As Phillipson and Powell (2004) critically suggest that 'Older people, it might be argued, are affected by two major changes: in respect of access to support on the one side, and the construction of identity on the other. On the one side, there is the creation of what Estes and others describe as "No Care Zones" where community supports may disintegrate in the face of inadequate services and benefits. On the other side, there may equally be the emergence of "No Identity Zones," these reflecting the absence of spaces in which to construct a viable identity for older people' (Phillipson & Powell, 2004: 55).

The lack of spaces adds to a sense of marginalization that exposes the vulnerable status of older people. But this vulnerability is not just about the material experience of deprivation; it also reaches into the texture of everyday life. For more affluent groups, a temporary solution seems to have been found in the denial of health and social care and the neo-liberal promotion of new consumer lifestyles. The social vacuum that this suggests reinforces the sense of uncertainty about the identity of older people in neo-liberal consumer society (Phillipson, 1998).

Chapter 10
Towards Global Health?

Introduction

Throughout this book, we have critically questioned taken-for-granted assumptions about health and social care in relation in particular to old age. The book aimed to provide a critical reflection to ideas and concepts of social theory and relevance to health and social care studies so as to facilitate understanding of modernist perspectives through to postmodern dimensions of human health. The aim of this chapter is to analyse the rapid expansion in the proportion of older people across the globe and to highlight the main social and economic forces causing this. Specific areas of the globe such as Americas, Europe, Asia and Africa will be focused on in detail before we discuss some of the key challenges and consequences of global aging for global society.

The chapter further highlights how globalisation and global aging are colossal driving forces that raise critical questions about the power of the individual nation state to deal with a global problem: a health population. Globalisation as both an analytical tool and social practice throws into flux the policies and practices of individual nation states to address social, economic and political issues for older people focusing on pensions and health and social care. It highlights how research needs to move from being state centred to one of which acknowledges global forces and the impact on populational health.

There is no doubt that the rapid increase in population health across the globe is signalling the most astonishing demographic changes in the history of humankind (Gruber & Wise, 2004). In every society in the world, there is concern about population health and its consequences for nation states, for sovereign governments and for individuals. The United Nations estimates that by the year 2025, the global population of those over 60 years will double, from 542 million in 1995 to around 1.2 billion people (Krug, 2002: 125). The global population age 65 or older was estimated at 461 million in 2004, an increase of 10.3 million just since

J. Powell, *New Perspectives on Health and Social Care*, International Perspectives on Social Policy, Administration, and Practice,
https://doi.org/10.1007/978-3-031-25432-1_10

2003. Projections suggest that the annual net gain will continue to exceed 10 million over the next decade: more than 850,000 each month. In 1990, 26 nations had older populations of at least 2 million, and by 2000, older populations in 31 countries had reached the 2 million mark (Cook & Powell, 2007). UN projections to 2030 indicate that more than 60 countries will have at least 2 million people age 65 or older.

While today's proportions of older people typically are highest in more developed countries, the most rapid increases in older populations are actually occurring in the less developed world (Krug, 2002). Between 2006 and 2030, the increasing number of older people in less developed countries is projected to escalate by 140% as compared to an increase of 51% in more developed countries (Krug, 2002). A key feature of population health is the progressive health of the older population itself. Over time, more older people survive to even more advanced ages. The forecast rise in the number of older people aged 75+ over the next 20 years will lead to an expansion of demand for health, housing accommodation and pensions for health populations and is thus of crucial importance for governments, policymakers, planners and researchers in all nation states. On a global level, the 85-and-over population is projected to increase 151% between 2005 and 2030, compared to a 104% increase for the population age 65 and over and a 21% increase for the population under age 65 (Bengston & Lowenstein, 2004). The most striking increase will occur in Japan: by 2030, nearly 24% of all older Japanese are expected to be at least 85 years old. As life expectancy increases and people aged 85 and over increase in number, four-generation families may become more common.

The age structure of the population has changed from one in which younger people predominated to a global society in which people in later life constituted a substantial proportion of the total population (Powell, 2005). Transformations in the age profile of a population are a response to political and economic structures. Older people in particular constitute a large section of populations in Western society in particular, but the percentage of pensionable age is projected to remain at 18% until 2011 when it becomes 20% and rising to 24% in 2025.

At the same time, there is a stigmatisation of such increasing populational numbers by ageist stereotypes. In relation to public services that have to be paid for by 'younger' working people, the percentage of the population has been used to signify such 'burdensome' numbers by the state (Estes et al., 2003). Dependency rates, that is, the number of dependants related to those of working age, have in fact altered little over the past 100 years. The reason there has been so little change during a period of so-called rapid health populations is that there has been a fall in the total fertility rate (the average number of children that would be born to each woman if the current age-specific birth rates persisted throughout her child-bearing life).

In advanced capitalist or First World countries, declines in fertility that began in the early 1900s have resulted in current fertility levels below the population replacement rate of two live births per woman. Perhaps the most surprising demographic development of the past 20 years has been the pace of fertility decline in many less developed countries (Giddens, 1993). In 2006, for example, the total fertility rate was at or below the replacement rate in 44 less developed countries (Cook & Powell, 2007). Most of the more developed nations have had decades to adjust to this change in age structure. For example, it took more than a century for France's population age 65 and over to increase from 7% to 14% of the total population. In contrast, many less developed or Third World countries are experiencing rapid increases in the number and percentage of older people, often within a single generation. The same demographic health process that unfolded over more than a century in France will occur in two decades in Brazil (OECD, 2007). In response to this compression of health, institutions must adapt quickly to accommodate a new age structure. Some less developed nations will be forced to confront issues, such as social support and the allocation of resources across generations, without the accompanying economic growth that characterised the experience of health societies in the West. In other words, some countries 'may grow old before they grow rich'.

Globalisation has also produced a distinctive stage in the social history of populational health, with a growing tension between nation state-based solutions (and anxieties) about growing old and those formulated by global institutions (Powell, 2005). Globalisation, defined here as the process whereby nation states are influenced (and sometimes undermined) by trans-national actors (Powell, 2005), has become an influential force in shaping responses to population health. Growing old has, itself, become relocated within a trans-national context, with international organisations (such as the World Bank and International Monetary Fund) and cross-border migrations, creating new conditions and environments for older people.

Aging can no longer just be viewed as a 'national' problem but one that affects trans-national agencies and communities. Local or national interpretations of aging had some meaning in a world where states were in control of their own destiny (Estes et al., 2003). They also carried force where social policies were being designed with the aim or aspiration of levelling inequalities, and where citizenship was still largely a national affair (and where there was some degree of confidence over what constituted 'national borders'). The crisis affecting each of these areas, largely set in motion by different aspects of globalisation, is now posing acute challenges for understanding 'global health' in the twenty-first century.

Global Health in Perspective

If these examples illustrate the complexity and impact of global aging, then it may be pertinent to highlight how populational health is impacting more specifically across different continents across the globe. The following section looks at health in four key areas across the globe: (i) Americas, (ii) Asia, (iii) Europe and (iv) Africa. These areas illustrate how population growth is impacting and creating social implications concerning health and disease as well as economic concerns relating to the labour market and pensions.

(i) Americas

Since the turn of the last century, the life expectancy of people born in North America has increased by approximately 25 years, and the proportion of persons 65 years or older has increased from 4% to over 13% (Estes & Associates, 2001). By the year 2030, one in five individuals in the United States is expected to be 65 years or older, and people age 85 and older make up the fastest growing segment of the population. In 2000, there were 34 million people aged 65 or older in the United States that represented 13% of the overall population (Estes & Associates, 2001). By 2030 there will be 70 million over 65 in the United States, more than twice their number in 2000. 31 million people, or 12% of the total population, are aged 65 and older. In another 35 years, the elderly population should double again. The health population is not only growing rapidly, but it is also getting older: 'In 1990, fewer than one in ten elderly persons was age 85 or older. By 2045, the oldest old will be in five. Increasing longevity and the steady movement of baby boomers into the oldest age group will drive this trend' (Longino, 1994: 856).

The percentage of oldest old will vary considerably from country to country. In the United States, for example, the oldest old accounted for 14% of all older people in 2005. By 2030, this percentage is unlikely to change because the health baby boom generation will continue to enter the ranks of the 65-and-over population (Bengston & Lowenstein, 2004). This is obviously causing much concern among policymakers, but Longino (1994), for instance, believes that thanks to better health, changing living arrangements and improved assistive devices, the future may not be as negative as we think when we consider an aging population.

It will be different, however, not least because people currently divorced constitute a small proportion of older populations. This will soon change in many countries as younger populations with higher rates of divorce and separation, age. In the United States, for example, 9% of the 65-and-over population is divorced or separated compared to 17% of people age 55–64 and 18% of people age 45–54 (Manton & Gu, 2001). This trend has gender-specific implications: in all probability, nonmarried women are less likely than nonmarried men to have accumulated assets and pension wealth for use in older age, while older men are less likely to form and maintain supportive social networks.

Shoring up public pensions is hardly the only avenue nations in North and South America are exploring. In many countries, privately managed savings accounts have been strongly advocated (Estes & Associates, 2001). Two decades ago, nearly every South American nation had pay-as-you-go systems similar to the US Social Security system. Some granted civil servants retiring in their 50s full salaries for life. Widening budget deficits changed that. In 1981, Chile replaced its public system with retirement accounts funded by worker contributions and managed by private firms. The World Bank encouraged 11 other Latin nations to introduce similar features. For example, in Chile the government addressed its fiscal budget deficit by mobilising a $49 billion of pension fund assets that make it easier for companies and corporations to fund investments in the local currency with bond offerings, and most workers have some retirement benefits from this (OECD, 2007). At the same time, the downside has been those people who cannot afford a private pension have been left to a low state pension which has intensified poverty (Estes & Associates, 2001), an enduring feature of all nation states in America. For the future, there is no safety guarantee that private pension schemes are protected and pay out for people who invest their savings in such provision. In a deregulated US pension system, the issue of corporate crime has highlighted the continuing problem of private pension provision. In one example, this was seen clearly with the energy corporation of Enron's embezzlement of billions of dollars of employees private pension schemes (Powell, 2005). This debate amounts to a significant global discourse about pension provision and retirement ages, but one which has largely excluded perspectives which might suggest an enlarged role for the state, and those which might question the stability and cost effectiveness of private schemes. The International Labour Organization (ILO) concluded that investing in financial markets is an uncertain and volatile business: that under present pension plans, people may save up to 30% more than they need, which would reduce their spending during their working life; or they may save 30% too little – which would severely cut their spending in retirement (Phillipson, 1998; Estes et al., 2003).

Holtzman (1997), in a paper outlining a World Bank perspective on pension reform, has argued for reducing state pay-as-you-go (PAYG) schemes to a minimal role of basic pension provision. This position has influenced both national governments and trans-national bodies, such as the International Labour Organization (ILO), with the latter now conceding to the World Bank's position with their advocacy of a mean-tested first pension, the promotion of an extended role for individualised and capitalised private pensions and the call for Organisation for Economic Co-operation and Development (OECD) member countries to raise the age of retirement.

There is also the impact of (intergovernmental organisations) IGOs on the pension debate in South America. The function of such arguments is to create a climate of fear, of inevitability and scientific certainty that public pension provision will fail. In so far as this strategy succeeds, it creates a self-fulfilling prophecy. If people believe the 'experts' who say publicly sponsored PAYG systems cannot be sustained, they are more likely to act in ways that mean they are unsustainable in practice. Certainly, in Europe and elsewhere, the state pension is an extremely popular institution. To have it removed or curtailed creates massive opposition. Only by demoralising the population with the belief that it is demographically unsustainable has room for the private financiers been created and a mass pensions market formed.

Increasingly, the social infrastructure of welfare states is being targeted as a major area of opportunity for global investors. The World Bank has expressed the belief that the public sector is less efficient in new infrastructure activities and that the time has come for private actors to provide what were once assumed to be public services. This view has been strongly endorsed by a variety of multinational companies, especially in their work with the World Trade Organization (WTO). The WTO enforces more than 20 separate international agreements, using international trade tribunals that adjudicate disputes. Such agreements include the General Agreement on Trade in Services (GATS), the first multilateral legally enforceable agreement covering banking, insurance, financial services and related areas (Estes et al., 2003).

(ii) Asia

Asia has the fastest increase in the aging population in the world. As we have cited in previous work, China in particular has been identified as having four 'unique characteristics' of populational health (Du & Tu, 2000).

1. Unprecedented speed: the proportion of health population is growing faster than Japan, the country previously recognised as having the fastest rate, and much faster than nations in Western Europe, for example.
2. Early arrival of a health population: before modernisation has fully taken place, with its welfare implications. 'It is certain that China will face a se-verely aged population before it has sufficient time and resources to estab-lish an adequate social security and service system for the elderly' (Du & Tu, 2000, 79).
3. Fluctuations in the total dependency ratio: the Chinese government estimates are that the country will reach a higher 'dependent burden' earlier in the twenty-first century than was previously forecasted.
4. Strong influence of the government's fertility policy and its implementation on the health process: the SCFP means fewer children being born, but with more elderly people, a conflict arises between the objectives to limit population increase and yet maintain a balanced age structure.

The combination of such factors means that the increased health population is giving rise to serious concerns among Chinese policymakers. Kim and Lee (2007) claim the growing elderly population is beginning to exert pressure on the East Asian countries' economies. Three decades ago, major industrialised countries have begun to grapple with the similar problem. With increasing drop in fertility rates, more East Asian economies such as Japan, Hong Kong, South Korea, Singapore and Taiwan are expected to turn into 'super-health societies' by 2025 (Kim & Lee, 2007). However, the magnitude of the future impact depends on the (in)ability of individual economies to resolve the demographic changes problem through increased privatisation, pension reforms, a migration on more productive countries and extension of retirement age. Like Western countries, Asia will ultimately have to tackle issues related to pension reform and the provision of long-term healthcare services (Cook & Powell, 2007).

For Japan, the basic statistical reality of its demographic profile is escalating. Already, 17 of every 100 of its people are over 65, and this ratio will near 30 in 15 years. From 2005 to 2012, Japan's workforce is projected to shrink by around 1% each year – a pace that will accelerate after that. Economists fear that, besides blowing an even bigger hole in Japan's underfunded pension system (Cook & Powell, 2007), the decline of workers and young families will make it harder for Japan to generate new wealth.

The future challenge of providing for the elderly is especially urgent in the world's two biggest nations – India and China. Only 11% of Indians have pensions, and they tend to be civil servants and the affluent. With a young population and relatively big families, many of the elderly population still count on their children for support. This is not the case in China. By 2030, there will be only two working-age people to support every retiree. Yet only 20% of workers have government- or company-funded pensions or medical coverage (Cook & Powell, 2007). However, as a counterbalance to such a gloomy perspective, 'Chindia' (China and India taken together) is currently accumulating vast wealth as a result of global change, wealth that could potentially be redirected for the support of their elderly populations.

(iii) Europe

The population structure of Western European countries has changed since the turn of the twentieth century. Whereas in 1901, just over 6% of the population were at or over current pension age (65 in the United Kingdom for men and women), this figure rose steadily to reach 18% in 2001 (Powell, 2005). At the same time, the population of younger people under age 16 fell from 35% to 20%. As European countries reach a relatively high level of population health, the proportion of workers tends to decline. European countries, including France, Germany, Greece, Italy, Russia and Ukraine, already have seen an absolute decline in the size of their workforce. And in countries where tax increases are needed to pay for transfers to growing older populations, the tax burden may discourage future workforce participation. The impact on a nation state's gross domestic product will depend on increases in labour productivity and that state's ability to substitute capital for labour. Less developed countries can shift their economies from labour-intensive to capital-intensive sectors as population health advances. Options for more European nation states may be more constrained. The 'rolling back' of pensions promises is just one symptom of a shift in European history: the 'graying of the baby-boom generation' (Phillipson, 1998). The percentage of 60-year-olds and older are growing 1.9% a year. This is 60% faster than the overall global population. In 1950 there were 12 people aged 15 to 64 to support each one of retirement age. Currently, the global average is nine. It will be only four-to-one by 2050 (Powell, 2005). By then numbers of older people will outnumber children for the first time. Some economists fear this will lead to bankrupt pensions and lower living standards. It is interesting that in Germany this fear is becoming a battleground for political electioneering. For example, Germany has the highest population in Europe and the third oldest population in the world, which presents both critical questions on public finances to provide pensions and healthcare and an opportunity for innovations in the marketplace. Currently, health has started to figure prominently in political discussions prior to 2009 elections, as political parties vie for the elderly vote. The current Merkel administration (2007–) has been criticised for increasing pensions, while opponents talk about a 'war of generations' requiring young people to pay for taxation for elder care.

The trend has drawn further attention across Europe, where the working-age population will decline by 0.6% in 2010. By 2025 the number of people aged 15 to 64 is projected to dwindle by 10.4% in Spain, 10.7% in Germany and 14.8% in Italy. But health is just as dramatic in such emerging markets as China – which is expected to have 265 million 65-year-olds by 2020 – and Russia and Ukraine (Cook & Powell, 2007).

Using evidence from the United Kingdom, the percentage of people of working age, that is, 16–64, will drop from 64% in 1994 to 58% in 2031 (Powell, 2005). As the number of workers per pensioner decreases, there will be pressure on pension provision. This is evident now, in such areas of pensions and long-term care, the re-treat of the state made evident in the erosion of State Earnings Related Pay are forcing people to devise their own strategies for economic survival in old age (Phillipson, 1998). In the British context that also impinges on global societies in general, private pensions are slowly being introduced in order to prevent the 'burden' of a health population. These are ways in which the state continues to rely on apocalyptic projections such as 'demographic time bomb' about health populations in order to justify cuts in public expenditure (Powell, 2005). Hence, the population of Great Britain, like that of other European countries, is aging rapidly. There are only enough young people to fill one in three of the new and replacement jobs that will need to be taken up over the next decade. Older people take much of the responsibility for our social and civic life and for the care of children, the sick and the very old in the community. Yet the gap between wealth and poverty, choice and the absence of choice for older people is stark and growing wider (Phillipson, 1998). The UK government is at the time of writing seeking to promote a debate over what they envisage as a multi-billion pound deficit that will be found in care for the elderly in the future.

(iv) Africa

Economic security, health and disability and living conditions in old age are policy concerns throughout the world, but the nature of the problem differs considerably from continent to continent and between and within countries – especially within Africa.

In Africa older people make up a relatively small fraction of the total population, and traditionally their main source of support has been the household and family, supplemented in many cases by other informal mechanisms, such as kinship networks and mutual aid societies. In 2005, Nigeria ranked among the top 30 countries in the world on the basis of the size of its population age 60 and over. Nigeria had the largest older population in sub-Saharan Africa, with over 6 million people age 60 and over; South Africa had just over 3.4 million. Congo and South Africa are projected to have nearly 5 million older people in 2030. Burkina Faso, Cameroon, Cote d'Ivoire, Madagascar, Mozambique, Niger, Senegal and Uganda are all projected to have their older populations grow to over one million people by 2030 (Building Blocks, 2004). Very little careful empirical research has been undertaken on long-term trends in the welfare of older people, but there are a number of reasons to believe that traditional caring and social support mechanisms in Africa are under increasing strain (OECD, 2007).

Located on the least developed and poorest continent, African economies are still heavily dependent on subsistence agriculture, and average income per capita is now lower than it was at the end of the 1960s. Consequently, the region contains a growing share of the world's poor. In addition, reductions in fertility and child mortality have meant that, despite the huge impact of the HIV/AIDS epidemic across much of the region, both the absolute size and the proportion of the population age 60 and over have grown and will continue to grow over the next 30 years (Estes et al., 2003).

In Africa, as in other traditional societies such as those in India or China, older people have traditionally been viewed in a positive light, as repositories of information and wisdom. And while African families are generally still intact, development and modernisation are closely connected with social and economic changes that can weaken traditional social values and networks that provide care and support in later life. Africa has long carried a high burden of disease, including from malaria and tuberculosis; today it is home to more than 60% of all people living with HIV, some 25.8 million in 2005. The vast majority of those affected are still in their prime wage-earning years, at an age when, normally, they would be expected to be the main wage earners and principal sources of financial and material support for older people and children in their families. Many older people have had to deal with the loss of their own support while absorbing the additional responsibilities of caring for their orphaned grandchildren. Increasingly, then, it appears that African societies are being asked to cope with population health with neither a comprehensive formal social security system nor a well-functioning traditional care system in place (Building Blocks, 2004).

The big issue is that majority of the world's population of older people (61%, or 355 million) live in poorer African countries. This proportion will increase to nearly 70% by 2025. For many countries, however, population aging has been accompanied by reductions in per capita income and declining living standards. Epstein (2001) notes that between 1950 and the late 1970s, life expectancy increased by at least 10% in every developing country in the world, or on average by about 15 years. However, at the beginning of the twenty-first century, life expectancy remains below 50 in more than 10 developing countries, and since 1970 has actually fallen, or has barely risen in a number of African countries (Phillipson, 1998). The AIDS epidemic is certainly a major factor here, but development loans requiring the privatisation of health care have also had an impact. Epstein (2001) reports, for example, that by the mid-1990s, the African continent was transferring four times more in debt repayment than it spent on health or education. More generally, HelpAge International (2000: 8) argues that 'Older people's poverty is still not a core concern in the social, economic and ethical debates of our time. Their right to development is routinely denied, with ageing seen as a minority interest or case for special pleading. Poverty and social exclusion remain the main stumbling blocks to the realisation of the human rights of older people worldwide'.

The Challenges and Consequences of Global Health

While global health represents a triumph of medical, social and economic advances, it also presents tremendous challenges. Population health strains social insurance and pension systems and challenges existing models of social support. It affects economic growth, trade, migration, disease patterns and prevalence and fundamental assumptions about growing older.

Older people's living arrangements reflect their need for family, community or institutional support. Living arrangements also indicate sociocultural preferences – for example, some choose to live in nuclear households, while others prefer extended families (Estes et al., 2003). The number, and often the percentage, of older people living alone is rising in most countries. In some European countries, more than 40% of women age 65 and older live alone. Even in societies with strong traditions of older parents living with children, such as in Japan, traditional living arrangements are becoming less common. In the past, living alone in older age often was equated with social isolation or family abandonment. However, research in many cultural settings illustrates that older people, even those living alone, prefer to be in their own homes and local communities (Gilleard & Higgs, 2001). This preference is reinforced by greater longevity, expanded social benefits, increased home ownership, elder-friendly housing and an emphasis in many nations on community care.

Global health will have dramatic effects on local, regional and global economies. Most significantly, financial expenditures, labour supply and total savings will be affected. Changes in the age structures of societies also affect total levels of labour force participation in society, because the likelihood that an individual will be in the labour force varies systematically by age. Concurrently, global population health is projected to lead to lower proportions of the population in the labour force in highly industrialised nations, threatening both productivity and the ability to support a health population (Krug, 2002).

Coupled with rapid growth in the young adult population in Third World countries, the World Bank foresees growing 'threats' to international stability pitting different demographic-economic regions against one another. The United Nations (2002) views the relationship between health populations and labour force participation with panic, recognising important policy challenges, including the need to reverse recent trends towards decreasing labour force participation of workers in late middle and old age despite mandatory retirement in Western countries such as the United Kingdom (Powell, 2005). Social welfare provisions and private sector pension policies influencing retirement income have a major impact on retirement timing. Hence, a major concern for organisations such as United Nations and World Bank centres on the number of such 'dependent' older people in all developing societies.

Some have argued that the rise of globalisation exerts unequal and highly stratified effects on the lives of older people (Estes & Associates, 2001). In the developed world, the magnitude and absolute size of expenditure on programmes for older people has made these the first to be targeted with financial cuts. In Third World countries, older people (women especially) have been among those most affected by the privatisation of health care, and the burden of debt repayments to the World Bank and the IMF (Estes & Associates, 2001). Additionally, globalisation as a process that stimulates population movement and migration may also produce changes that disrupt the lives of older people. And one must not forget either that they may comprise up to one third of refugees in conflict and emergency situations – a figure which was estimated at over 53 million older people worldwide in 2000 (Estes & Associates, 2001).

Nation states with extensive social programmes targeted to the older population (principally healthcare and income support programmes) find the costs of these programmes escalating as the number of eligible recipients grows and the duration of eligibility lengthens due to global pressures. Further, few countries have fully funded programmes; most countries fund these programmes on a pay-as-you-go basis or finance them using general revenue streams. Governments may be limited in how much they can reshape social insurance programmes by raising the age of eligibility, increasing contribution rates and reducing benefits. Consequently, shortfalls may need to be financed using general revenues. Projections of government expenditures in the United States and other OECD countries show increases in the share of gross domestic product devoted to social entitlements for older populations. In some cases, this share more than doubles as a result of population health (OECD, 2007).

Different countries' age groups have different levels of pace of growth. It is possible for the elements of production – labour and capital – to flow across national boundaries and mitigate the impact of population aging. Studies predict that, in the near term, surplus capital will flow from Europe and North America to emerging markets in Asia and Latin America, where the population is younger and supplies of capital relatively low. In another 20 years, when the baby boom generation in the West has mostly retired, capital likely will flow in the opposite direction. Traditionally, labour is viewed as less mobile than capital, although migration could offset partially the effects of population health. Currently, 22% of physicians and 12% of nurses in the United States are foreign born, representing primarily African countries, the Caribbean and Southeast Asia (OECD, 2007). The foreign-born workforce also is growing in most OECD countries. Over the next 10 years, the European experience will be particularly instructive in terms of the interplay of health and migration (OECD, 2007). Some pressure groups are now suggesting, for instance, that a rich city like London, which benefits from Ghanaian nurses in the National Health Service, has an ethical obligation to Ghana itself, to provide funds to support that country's health training system because the donor country is losing key personnel.

The life-cycle theory of consumption is that households accumulate wealth during working years to maintain consumption in retirement (Gilleard & Higgs, 2001). The total of a country's individual life-cycle savings profiles determines whether households in that country are net savers or nonsavers at any point in time. A country with a high proportion of workers will tend to be dominated by savers, placing downward pressure on the rate of return to capital in that economy. Nation states with older populations will be tapping their savings and driving rates of return higher because of the scarcity of capital (Gilleard & Higgs, 2001).

Retirement resources typically include public and private pensions, financial assets and property. The relative importance of these resources varies across countries. For example, a groundbreaking study revealed that only 3% of Spanish households with at least one member age 50 or older own stocks (shares), compared to 38% of Swedish households. The largest component of household wealth in many countries is housing value. This value could fall if large numbers of older homeowners try to sell houses to smaller numbers of younger buyers. How successfully this transition is managed around the world could determine the rise and fall of nations and reshape the global economy in the era of the post-credit crunch. Two key vehicles of growth are increases in the labour force and productivity. If nation states cannot maintain the size of their labour forces by persuading older workers to retire later, then the challenge will be to maintain growth levels. That will be a particular challenge in Europe, where productivity growth has averaged just 1.3% since 1995. By 2024, growth in household financial wealth in the United States, Europe and Japan will slow from a combined 4.5% annual reduction now to 1.3%. That will translate into $31 trillion less wealth than if the average age were to remain the same (Cook & Powell, 2007).

Most of Europe's state-funded pension systems encourage early retirement. Now, 85.5% of adults in France retire from employment by age 60, and only 1.3% engage in employment beyond aged 65. In Italy, 62% of adults retire from full-time work by the age of 55. That compares with 47% of people who earn wages or salaries until they are 65 in the United States and 55% in Japan (Estes & Associates, 2001).

Why the sudden attention to a demographic trend of global health? In part, it is because the future is already dawning that global trends impact on state power. In South Korea and Japan, which have strong cultural aversions to immigration, small factories, construction companies and health clinics are relying more on 'temporary' workers from the Philippines, Bangladesh and Vietnam (OECD, 2007). In China, state industries are struggling over how to lay off unneeded middle-age workers when there is no social safety net to support them.

What really has pushed health to the top of the global agenda, though, are increasing fiscal gaps in part, due to the 'global credit crunch' in the United States, Europe, Japan and elsewhere that could worsen as populations reach retirement age. While US Social Security is projected to remain solvent until at least 2042, the picture is more acute in Europe. Unlike the United States, where most citizens also have private savings plans, in much of Europe, up to 90% of workers rely almost entirely on public pensions. Austria guarantees 93% of pay at retirement, for example, and Spain offers 94.7%. Pensions and elder-care costs will increase from 14% of capitalist nations' gross domestic product to 18% by 2050.

As people live longer and have fewer children, family structures are also transformed (Bengston & Lowenstein, 2004). This has important implications in terms of providing care to older people. Most older people today have children, and many have grandchildren and siblings. However, in countries with very low birth rates, future generations will have few if any siblings. As a result of this trend and the global trend towards having fewer children, people will have less familial care and support as they age (Bengston & Lowenstein, 2004). Unless there is a fundamental shift in the views of 'Fortress Europe', Japan and other countries towards immigrants, and an overcoming of entrenched racial or racist attitudes towards migrants, some parts of the globe will be 'elderly heavy', while others will be 'elderly light'. Were migrants made more welcome in richer societies, then one could envisage a space of carer flows, with more interactions and movements in either direction to the 'heavy' or 'light' end. Or, for example, one could have elderly relocation in the same way as Japanese elderly are relocating into Thailand into new forms of 'trans-national households', in order to seek cheaper care systems for their retirement.

As a consequence of the global demographics of health and social care, the changing societies of the post millennium are being confronted with quite profound issues relating to illness and healthcare, access to housing and economic resources including pension provision. The past several years has witnessed an unprecedented stretching of the human life span. This health of the global population is without parallel in human history. If these demographic trends continue to escalate by 2050, the number of older people globally will exceed the number of young for the first time since formal records began raising questions of the power of the nation state in the context of global aging, and raising further global questions of distribution of power and scarcity of resources to an aging population in relation to fundamental issues associated with health and social care.

References

Acker, J. (1988). Class, gender, and the relations of distribution. *Signs., 13*(3), 473–297.

Adorno, T., & Horkheimer, M. (1944). *The culture industry: Enlightenment as mass deception; dialectic of enlightenment*. Continuum.

Allen, I., et al. (1992). *Elderly people: Choice, participation and satisfaction*. Policy Studies Institute.

Althusser, L. (1971). Ideology and ideological apparatuses (notes towards an investigation) (reprinted). In V. Beechey & J. Donald (Eds.), *Subjectivity and social relations* (1985). Open University Press.

Arber, S., & Ginn, J. (1991). *Gender and later life: A sociological analysis of resources and constraints*. Sage.

Arber, S., & Ginn, J. (Eds.). (1995). *Connecting gender and ageing: A sociological approach*. OUP.

Armstrong, D. (1983). *Political anatomy of the body: Medical knowledge in Britain in the twentieth century*. Cambridge University Press.

Austin, J. L. (1962). *How to do things with words*. Oxford University Press.

Baltes, M., & Carlestensen, L. (1996). The process of successful ageing. *Ageing & Society, 16*, 397–422.

Bauman, Z. (1992). *Imitations of postmodernity*. Routledge.

Beck, U. (1992). *Risk society: Towards a new modernity*. Routledge.

Beck, U. (1994). *The risk society*. Sage.

Beck, U. (1994). *Reflexive modernization*. Stanford University Press.

Beck, U. (1996). *The reinvention of politics*. Polity.

Beck, U. (1998). *Democracy without enemies*. Polity.

Beck, U. (2005). *Power and countervailing power in the global age*. Polity.

Beechey, V., & Donald, J. (1985). Introduction. In V. Beechey & J. Donald (Eds.), *Subjectivity and social relations*. Open University Press.

Bengtson, V., & Schaie, R. (1999). *Handbook of theories in gerontology*. Springer.

Bengston, V. L., & Lowenstein, A. (Eds.). (2004). *Global aging and challenges to families*. De Gruyter.

Bengtson, V. L., & Achenbaum, W. (1993). *The changing contract across generations*. Aldine De Gruyter.

Bengston, V., et al. (1997). Theory, explanation, and a third generation of theoretical development in social gerontology. *Journal of Gerontology., 52*(2), 72–88.

Bengston, V., et al. (2009). Older parent–child relationships in six developed nations: Comparisons at the intersection of affection and conflict. *Journal of Marriage and Family., 72*(4), 1006–1021.

J. Powell, *New Perspectives on Health and Social Care*, International Perspectives on Social Policy, Administration, and Practice,
https://doi.org/10.1007/978-3-031-25432-1

Bengtson, V. L., Giarrusso, R., Silverstein, M., & Wang, H. (2000). Families and intergenerational relationships in aging societies. *Hallym International Journal of Aging, 2*(1), 3–10.

Beresford, P., Croft, S., Evans, C., & Harding, T. (1997). Quality in personal social services: The developing role of user involvement in the U.K. In A. Evers, R. Haverinen, K. Leichsenring, & G. Winstow (Eds.), *Developing quality in personal social services: A European perspective*. Ashgate.

Bernauer, J. W., & Mahon, M. (1994). The ethics of Michel Foucault. In G. Gutting (Ed.), *The Cambridge companion to Foucault*. Cambridge University Press.

Better Government of Older People (BGOP). (2000). *Better government for career people*. BGOP.

Beveridge, W. (1942). *The Beveridge report*. HMSO.

Biggs, S. (1993). *Understanding Ageing*. OUP.

Biggs, S. (1996). A family concern: Elder abuse in British social policy. *Critical Social Policy, 16*(2), 63–88.

Biggs, S. (1999). *The mature imagination*. OUP.

Biggs, S. (2001). Toward critical narrativity: Stories of aging in contemporary policy. *Journal of Aging Studies, 15*, 1–14.

Biggs, S., & Powell, J. L. (1999). Surveillance and elder abuse: The rationalities and technologies of community care. *Journal of Contemporary Health, 4*(1), 43–49.

Biggs, S., & Powell, J. L. (2000). Surveillance and elder abuse: The rationalities and technologies of community care. *Journal of Contemporary Health, 4*(1), 43–49.

Biggs, S., & Powell, J. (2001). A Foucauldian analysis of old age and the power of social welfare. *Journal of Aging & Social Policy, 12*(2), 93–112.

Biggs, S., Phillipson, C., & Kingston, P. (1995). *Elder abuse in perspective*. Open University Press.

Biggs, S., Estes, C., & Phillipson, C. (2003). *Social theory, social policy and ageing*. Open University Press.

Blaikie, A. (1999). *Ageing and popular culture*. OUP.

Blair, T. (1996). *New Britain: My vision of a young country*. London.

Boaz, A., Hayden, C., & Bernard, M. (1999). *Attitudes and aspirations of older people* (DSS Research Report, No. 101). CDS.

Bond, J., & Coleman, P. (Eds.). (1993). *Ageing in society: An introduction to social gerontology*. Sage.

Bornat, J., Dimmock, B., Jones, D., & Peace, S. (1999). Stepfamilies and older people. *Ageing and Society, 19*(2), 239–262.

Bromley, D. (1966). *The psychology of human ageing*. Penguin.

Building Blocks. (2004). *Africa-wide briefing notes – Supporting older carers*. HIV AIDS Alliance and Help Age International.

Burchell, G. (1993). Liberal government and techniques of the self. *Economy and Society, 22*(3), 267–282.

Bury, M. (1995). Ageing, gender and sociological theory. In S. Arber & J. Ginn (Eds.), *Connecting gender and Ageing: A sociological approach*. OUP.

Butler, J. (1990). *Gender trouble: Feminism and the subversion of identity*. Routledge.

Butler, J. (1993). *Bodies that matter: On the discursive limits of 'sex'*. Routledge.

Butler, J. (1995). Burning acts, injurious speech. In A. Parker & E. K. Sedgwick (Eds.), *Performativity and performance*. Routledge.

Butler, J. (1998a). *The psychic life of power: Theories in subjection*. Stanford University Press.

Butler, J. (1998b). Merely cultural. *New Left Review, 227*, 33–34.

Bytheway, W. (1995). *Ageism*. OUP.

Callaghan, J. (1989). Case management for the elderly; a panacea? *Journal of Aging and Social Policy, 1*(2), 181–195.

Callinicos, A. (1990). *Against post-modernism*. Polity Press.

Carmel, S., Morse, C. A., & Torres-Gil, F. M. (Eds.). (2007). *Lessons on ageing from three nations*. Baywood.

Castel, R. (1991). From dangerousness to risk. In G. Burchell, C. Gordon, & P. Miller (Eds.), *The Foucault effect: Studies in governmentality*. University of Chicago Press.

Chamberlayne, P., & King, A. (2000). *Cultures of care*. Policy Press.
Chambre, S. M. (1993). Volunteerism by elders: past traditions and future prospects. *The Gerontologist, 33*, 221–228.
Chau, W. F. (1995). Experts, networks and inscriptions in the fabrication of accounting images. *Accounting Organisations and Society, 20*(2/3), 111–145.
Chudacoff, H. (1989). *How old are you?* Princeton University Press.
Clarke, J. (1994, December). *Capturing the customer: Consumerism and social welfare*. Paper to ESRC seminar conceptualising consumption issues, University of Lancaster.
Clarke, J., & Newman, J. (1997). *The managerial state*. Sage.
Clarke, J., et al. (1978). *Policing the crisis*. Open University Press.
Cloke, P., Johnsen, S., & May, J. (2006). Ethical citizenship? Volunteers and the ethics of providing services for homeless people. *Geoforum, 38*(6), 1089–1101.
Cole, T., Van Tassel, D., & Kastenbaum, R. (1992). *Handbook of the humanities and ageing*. Springer.
Cook, I. G., & Powell, J. L. (Eds.). (2007). *New perspectives on China and aging*. Nova Science.
Cousins, M., & Hussain, A. (1984). *Michel Foucault*. Macmillan.
Cumming, E., & Henry, W. (1961). *Growing old: The process of disengagement*. Basic Books.
Davidson, A. (1986). Archaeology, genealogy, ethics. In D. Hoy (Ed.), *Foucault: A critical reader*. Basil Blackwell.
Davidson, A. I. (1994). Ethics as ascetics: Foucault, the history of ethics, and ancient thought. In G. Gutting (Ed.), *The Cambridge companion to Foucault* (pp. 115–140). Cambridge University Press.
De Beauvoir, S. (1979). *Ageing*. Penguin.
Dean, M. (2007). *Governing societies*. Open University Press/McGraw-Hill.
Delanty, G. (1999). *Social theory in a changing world: Conceptions of modernity*. Polity Press.
Department of Health. (1989). *Caring for people: Community care in the next decade and beyond* (cm849). HMSO.
Department of Health. (1991). *Caring for people: Implementation documents, draft guidance: Assessment and care management*. HMSO.
Department of Health. (1996). *Community Care (Direct Payments) Act*. TSO.
Department of Health. (2004). *Every child matters: Change for children*. The Stationery Office.
Department of Health. (2005). *Independence wellbeing and choice: Our vision for the future of social care for adults in England* (Cm 6499). TSO.
Department of Health. (2006a). *Reward and recognition: The principles and practice of service user payment and reimbursement in health and social care: A guide for service providers, service users and carers*. The Stationery Office.
Department of Health. (2006b). *Our health, our care, our say: A new direction for community services* (CM 6737). The Stationery Office.
Department of Health. (2007). *Putting people first: Shared vision and commitment to the transformation of adult social care*. TSO.
Department of Health. (2010). *A vision for adult social care: Capable communities and active citizens*. TSO.
Department of Health. (2011). *Think local, act personal: A sector wide commitment to moving forward with personalisation and community based support*. TSO.
Department of Health and Home Office. (2000). *No secrets: Guidance on developing and implementing multi-health and social carency policies and procedures to protect vulnerable adults from abuse*. TSO.
Department of Health Security (DSS). (1998). *Building a better Britain for career people*. HMSO.
Derrida, J. (1988). *Signature, event, context*. Limited, Inc./North Western University Press.
Dews, P. (1989). The return of the subject in late Foucault. *Radical Philosophy, 51*, 37–41.
Dickinson, H., & Glasby, J. (2010). *The personalization health and social care agenda: Implications for the third sector* (Third Sector Research Centre: Working Paper 30). TSRC.
Dittman-Kolhi, F. (1991). The construction of meaning in ageing. *Ageing & Society, 10*, 279–294.

Donzelot, J. (1980). The policing of families. In *Foucault, the family and politics*. https://doi.org/10.1057/9781137291288_6

Dreyfus, H., & Rabinow, P. (1982a). *Michel Foucault: Beyond structuralism and hermeneutics*. University of Chicago Press.

Dreyfus, H. L., & Rabinow, P. (1982b). *Michel Foucault: Beyond structuralism and hermeneutics*. Harvester.

Dreyfus, H., & Rabinow, P. (1983). *Michel Foucault: Beyond structuralism and hermeneutics*. University of Chicago Pres.

Du, P., & Tu, P. (2000). Population ageing and old age security. In X. Peng & Z. Guo (Eds.), *The changing population of China* (pp. 77–90). Blackwell.

Dyer, J., & Keller-Cohen, D. (2000). The discursive construction of professional self through narratives of personal experience. *Discourse Studies, 2*(3), 283–304.

Elders, F. (Ed.). (1974). *Reflexive waters: The basic concerns of mankind*. Souvenir Press.

Epstein, H. (2001, April). Time of indifference. *New York Review of Books, 12*, 33–38.

Estes, C. (1979). *The aging enterprise*. Jossey Bass.

Estes, C., & Binney, E. (1989). The biomedicalization of aging. *The Gerontologist, 29*(5), 587–598.

Estes, C., & Associates. (2001). *Social policy and aging*. Sage.

Estes, C., Swan, J., & Gerard, L. (1982). Dominant and competing paradigms in old age: Towards a political economy of old age. *Ageing and Society, 12*, 151–164.

Estes, C., Biggs, S., & Phillipson, C. (2003). *Social theory, social policy and ageing*. Open University Press.

Ewald, F. (1993). *L'état providence*. Grasset Press.

Fairclough, N. (1992). *Discourse and social change*. Polity Press.

Fattah, E., & Sacco, V. (1989). *Crime and victimization of the elderly*. Springer.

Featherstone, M., & Hepworth, M. (1989). Ageing and old age: Reflections on the post-modern life course. In B. Bytheway (Ed.), *Becoming and being old: Sociological approaches to later life* (pp. 143–157). Sage.

Featherstone, M., & Hepworth, M. (1993). Images of ageing. In J. Bond, P. Coleman, & S. Peace (Eds.), *Ageing in society*. Open University Press.

Featherstone, M., & Wernick, A. (1995). *Images of Ageing*. Routledge.

Ferguson, I. (2007). Increasing user choice or privatizing risk? The antinomies of personalization. *British Journal of Social Work, 37*(3), 387–403.

Finch, J., & Mason, J. (1993). *Negotiating family responsibilities*. Routledge.

Fink, J., Lewis, G., Carabine, J., & Newman, J. (2004). *Personal lives and social policy: Course companion*. Sage.

Fleming, P. (2005). Metaphors of resistance. *Management Communication Quarterly, 19*(1), 45–66.

Fleming, P., & Sewell, G. (2002). Looking for the good soldier, Svejk; alternative modalities of resistance in the contemporary workplace. *Sociology, 36*(4), 857–872.

Flynn, R. (1992). *Structures of control in health management*. Routledge.

Forbat, L., & Atkinson, D. (2005). Advocacy in practice: The troubled position of advocates in adult services. *British Journal of Social Work, 35*, 321–335.

Foucault, M. (1965). *Madness and civilization*. Tavistock.

Foucault, M. (1967). *Madness and civilisation*. Tavistock.

Foucault, M. (1972). *The archaeology of knowledge*. Tavistock.

Foucault, M. (1973). *The birth of the clinic*. Routledge.

Foucault, M. (1976). *The history of sexuality*. Penguin.

Foucault, M. (1977a). *Discipline and punish*. Tavistock.

Foucault, M. (1977b). *Discipline and punish*. Penguin.

Foucault, M. (1977c). *Discipline and punish*. Allen Lane.

Foucault, M. (1978a). *The history of sexuality* (Vol. 1). Penguin.

Foucault, M. (1978b). *The history of sexuality: An introduction*. Penguin.

Foucault, M. (1978c). Governmentality. In G. Burchell (Ed.) (1991). *The Foucault effect: Studies in governmentality*. Harvester.

Foucault, M. (1978d). *The history of sexuality [Volume 1]: An introduction*. Vintage Books.

Foucault, M. (1980a). *Power/knowledge: Selected interviews and other writings, 1972–1977*. Pantheon.

Foucault, M. (1980b). *Power/knowledge: Selected interviews and other writings, 1972–1977* (C. Gordon, Ed.). Pantheon.

Foucault, M. (1982). The subject of power. In H. Dreyfus & P. Rabinow (Eds.), *Michel Foucault: Beyond structuralism and hermeneutics*. Harvester.

Foucault, M. (1983). The subject of power. In H. Dreyfus & P. Rabinow (Eds.), *Michel Foucault: Beyond structuralism and hermeneutics*. Brighton.

Foucault, M. (1984a). Space, knowledge, power. In P. Rabinow (Ed.), *The Foucault reader*. Penguin.

Foucault, M. (1984b). What is enlightenment? In P. Rabinow (Ed.), *The Foucault reader*. Peregrine.

Foucault, M. (1984c). *The Foucault reader* (P. Rabinow, Ed.). Penguin.

Foucault, M. (1985). Discourse and truth: The problematization of parrhesia. In J. Pearson (Ed.), *Foucault*. Northwestern University Press.

Foucault, M. (1988). Technologies of the self. In L. H. Martin et al. (Eds.), *Technologies of the self*. Tavistock.

Foucault, M. (1989). *Foucault live: Collected interviews 1961–1984* (E. Lotringer, Ed.; J. Johnston, Trans.). Semiotext(e).

Foucault, M. (1991a). *Remarks on Marx: Conversations with Duccio Trombadori* (R. J. Goldstein & J. Cascaito, Trans.). Semiotext(e).

Foucault, M. (1991b). The ethic of care for the self as a practice of freedom: An interview with Fornet-Betancourt, R., Becker, H. and Gomez-Müller, A. Translated by Gauthier Snr, J. D. In J. Bernauer & D. Rasmussen (Eds.), *The final Foucault*. MIT Press.

Foucault, M. (1991c). Governmentality. In G. Burchell (Ed.), *The Foucault effect: Studies in governmentality*. University of Chicago Press.

Foucault, M. (1991d). Governmentality. In G. Burchell, C. Gordon, & P. Miller (Eds.), *The Foucault effect: Studies in governmentality*. Harvester Wheatsheaf.

Fournier, V. (1999). The appeal to 'Professionalism' as a disciplinary mechanism. *The Sociological Review, 47*(2), 280–307.

Frank, A. (1998). Stories of illness as care of the self: A Foucauldian dialogue. *Health, 2*(3), 329–348.

Freidenberg, J. (2000). *Growing old in EL Barrio*. New York University Press.

George, L. K. (1995). Depression disorders and symptoms in later life. *Generations, 17*(1), 35–38.

Giddens, A. (1990). *The consequences of modernity*. Polity Press.

Giddens, A. (1991). *The consequences of modernity*. Polity.

Giddens, A. (1993). *Sociology*. Polity Press.

Giddens, A. (1998a). *The third way*. Polity.

Giddens, A. (1998b). *The third way: The renewal of social democracy*. Polity Press.

Gilbert, T. (2001). Reflective practice and clinical supervision: Meticulous rituals of the confessional. *Journal of Advanced Nursing, 36*(2), 199–205.

Gilbert, T., & Powell, J. L. (2005). Family, caring and aging in the UK. *Scandinavian Journal of Caring Sciences, 41*(2), 41–48.

Gilbert, T., & Powell, J. L. (2010). Power and social work in the United Kingdom: A Foucauldian excursion. *Journal of Social Work, 10*(1), 3–22.

Gilleard, C., & Higgs, P. (2000). *Cultures of ageing*. Prentice Hall.

Gilleard, C., & Higgs, P. (2001). *Cultures of aging*. Prentice Hall.

Gilleard, C., & Higgs, P. (2005). *Contexts of ageing: Class, cohort and community*. Polity Press.

Glendinning, C., Challis, D. J., Fernandez, J. L., Jacobs, S., Jones, K., Knapp, M., Manthorpe, J., Moran, N., Netten, A., Stevens, M., & Wilberforce, M. (2008). *Evaluation of the individual budgets pilot programme: Final report*. Social Policy Research Unit, University of York.

Goffman, E. (1959). *The presentation of self in everyday life*. Penguin.

Goffman, E. (1968) *Total institutions*. Routledge.

Gorbien, M. (2011). *Protecting against fraud and financial abuse*. Rush University Medical Center.

Gregson, N., & Rose, G. (2000). Taking Butler elsewhere: Performativities, spatialities and subjectivities. *Environment and Planning D: Society and Space, 18*(4), 433–452.

Grey, C. (1998). On being A professional in a 'Big Six' firm. *Accounting, Organizations and Society, 23*(5–6), 569–587.

Gruber, J., & Wise, D. A. (Eds.). (2004). *Social security programs and retirement around the world. Micro estimation*. University of Chicago Press.

Gubrium, J. (1975). *Out of control: Family therapy and domestic disorder*. Sage.

Gubrium, J. F. (1992). *Out of control: Family therapy and domestic disorder*. Sage.

Gubrium, J. F. (1998). Narrative practice and the coherence of personal stories. *The Sociological Quarterly, 39*(1), 163–187.

Gutting, G. (2003). *The Cambridge companion to Foucault*. Cambridge University Press.

Habermas, J. (1992). *Postmetaphysical thinking*. Polity.

Hacking, I. (1990). *The taming of chance*. Cambridge University Press.

Haraway, D. (1991). *Simians, cyborgs and women*. Routledge.

Havighurst, R. J., & Albrecht, R. (1953). *Older people*. Longman.

Hall, S. (2001). The spectacle of the other. In M. Wetherell, S. Taylor, & S. J. Yates (Eds.), *Discourse theory and practice*. Sage.

Hardill, I., Baines, S., & 6, P. (2007). Volunteering for all? Explaining patterns of volunteering and identifying strategies to promote it. *Policy and Politics, 35*(3), 395–412.

Hargrave, T., & Anderson, W. (1992). *Finishing well: Health and social care and reparation in the intergenerational family*. Brunner and Mazel.

Hayden, C., Boaz, A., & Taylor, F. (1999). *Attitudes and aspirations of career people: A qualitative study* (DSS Research Report, No. 102). CDS.

Healy, L. (2001). *International social work: Professional action in a interdependent world*. OUP.

Henwood, M. (1995). *Making a difference?: Implementation of the community care reforms two years on*. Kings Fund Press.

Help Age International. (2000). *The mark of a noble society*. Help Age International.

Hochschild, A. R. (1983). *The managed heart: The commercialization of human feeling*. University of California Press.

Hogg, D., & Peace, S. (1992). *Elderly people: choice, participation and satisfaction*. Policy Studies Institute.

Holstein, J., & Gubrium, J. (2000). *The self we live by*. Oxford University Press.

Hollingdale, R. (1977). *The Nietzsche reader*. Penguin.

Honneth, A., & Joas, H. (1988). *Social action and human nature*. Cambridge University Press.

Holtzman, R. A. (1997, December). *A World Bank perspective on pension reform*. Paper prepared for the joint ILO-OECD workshop on the development and reform of pension schemes, Paris.

Howe, D. (1994). Modernity, postmodernity and social work. *British Journal of Social Work., 24*(5), 513–532.

Hoyle, L., & Le Grand, J. (1991). *Markets in social care services*. SAUS.

Hughes, B. (1995). *Older people and community care*. OU Press.

Ignatieff, M. (1978). *A just measure of pain: The penitentiary in the industrial revolution, 1750–1850*. Pantheon Books.

Irvine, E. (1954). Research into problem families. *British Journal of Psychiatric Social Work, No., 9*(1), 110–118.

Izuhara, M. (2000). *Family change and housing in postwar Japanese society*. Ashgate.

Jessop, B. (1994). The transition to post-Fordism and the Schumpeterian workfare state. In R. Burrows & B. Loader (Eds.), *Towards a post-Fordist welfare state?* Routledge.

Johnson, T. (2001). Governmentality and the institutionalization of expertise. In M. Purdy & D. Banks (Eds.), *The sociology and politics of health: A reader* (pp. 135–143). Routledge.

Jones, C. (1983). *State social work and the working class*. Macmillan.

Jones, C. (1983). Social work today. *The British Journal of Social Work., 23*(3), 195–212.

Jordon, B. (2005). New labour: Choice and values. *Critical Social Policy, 25*(4), 427–446.

Kastenbaum, R. (1993). Encrusted elders. In T. Cole (Ed.), *Voices & visions of health and social care*. Springer.
Katz, S. (1996). *Disciplining old age: The formation of gerontological knowledge*. University of Virginia.
Katz, S. (1998). *Lifecourse, lifestyle and postmodern culture: Corporate representations of later life*. Paper presented at restructuring work and the lifecourse: An international symposium, Institute of Human Development, University of Toronto.
Katz, S. (1999). Busy bodies: Activity, aging and the management of everyday life. *Journal of Aging Studies, 14*(2), 135–152.
Katz, S. (2007). The problems of eurocentrism and evolutionism in Marx's writings on colonialism. *Political Studies., 28*(4), 672–686.
Kelly, C. (1994). Social identity and intergroup perceptions in minority-majority contexts. *Human Relations, 43*(2), 583–599.
Kendall, J. (2003). *The voluntary sector: Comparative perspectives in the UK*. Routledge.
Kennedy, G. (1990). College students expectations of grandparent and grandchild role behaviours. *The Gerontologist, 30*(1), 43–48.
Kenyon, G., Ruth, J., & Mader, W. (1999). Elements of a narrative health and social care. In V. Bengtson & K. Schaie (Eds.), *Handbook of theories of health and social careing*. Springer.
Kim, S., & Lee, J.-W. (2007). Demographic changes, saving and current account in East Asia. *Asian Economic Papers, 6*(2), 22–53.
Kivnick, H. (1988). Grandparenthood, life review and psychohealth development. *Journal of Gerontological Health Work, 12*(3), 63–82.
Knowles, C. (2001). Cultural perspectives and welfare regimes: The contributions of Foucault and Lefebve. In P. Chamberlayne, A. Cooper, & M. Rustin (Eds.), *Welfare and culture in Europe: Towards a new paradigm in social policy* (pp. 240–254). Jessica Kingsley.
Krug, E. G. (2002). *World report on violence and health*. World Health Organization.
Land, H. (1999). *The changing worlds of work and families*. Open University Press.
Layder, D. (2006). *Understanding social theory*. Sage.
Leonard, P. (1997). *Postmodern welfare*. Sage.
Lewis, J., & Glennerster, H. (1996). *Implementing the new community care*. OUP.
Longino, C. F. (1994). Pressure from our aging population will broaden our under-standing of medicine. *Academic Medicine, 72*(10), 841–847.
Longino, C. F., & Powell, J. L. (2004). Embodiment and the study of aging. In V. Berdayes (Ed.), *The body in human inquiry*. The Hampton Press.
Lubove, R. (1966, May 23). Social work and the life of the poor. *The Nation*, pp. 609–611.
Luhmann, N. (1993). *Risk: A sociological theory*. De Gruyter.
Lupton, D. (Ed.). (1999). *Risk and sociocultural theory: New directions and perspectives*. Cambridge University Press.
Manthorpe, J., & Stevens, M. (2010). Understanding the potential impact of personalization for social work with rural career people. *British Journal of Social Work, 40*(5), 1452–1469.
Manton, K. G., & Gu, X. (2001). Changes in the prevalence of chronic disability in the United States black and nonblack population above age 65 from 1982 to 1999. *Proceedings of the National Academy of Sciences, 98*, 6354–6359.
Massumi, B. (1993). *The politics of everyday fear*. Routledge.
May, T. (1991). *Probation: Politics, policy and practice*. Open University Press.
May, T. (1994). Transformative power: A study in a human service organisation. *The Sociological Review, 42*(4), 618–638.
May, T. (1999). From Banana time to just-in-time: Power and resistance at work. *Sociology, 33*(4), 767–783.
May, T., & Powell, J. (2007). Michel Foucault. In T. Edwards (Ed.), *Cultural theory: Classical and contemporary positions*. SAGE.
McAdams, D. (1993). *The stories we live by*. Morrow.
McLeod, J. (1997). *Narrative and psychotherapy*. Routledge.

McNay, L. (1992). *Foucault and feminism: Power, gender and the self.* Polity Press.
McNay, L. (1993). *Foucault and feminism.* Polity.
McNay, L. (1994). *Foucault.* Cambridge.
Mental Capacity Act (2005). TSO.
Miehis, D., & Moffatt, K. (2000). Constructing social work identity based on the reflexive self. *British Journal of Social Work, 30*, 339–348.
Miller, T. (1993). *The well-tempered self: Citizenship, culture and the postmodern subject.* John Hopkins University Press.
Mills, C. W. (1959). *The sociological imagination.* Oxford University Press.
Mills, T. (1999). When grandchildren grow up. *Journal of Aging Studies, 13*(2), 219–239.
Milner, J., & O'Byrne, P. (1998). *Assessment in social work.* Macmillan.
Minkler, M. (1999). Intergenerational househcares headed by grandparents: con-texts, realities and implications for policy. *Journal of Health and Social Care Studies, 13*(2), 199–218.
Mölling, G. (2001). The nature of trust: From George Simmel to a theory of expectation, interpretation and suspension. *Sociology, 35*(2), 403–420.
Moody, H. (1998). *Aging, concepts and controversies.* Pie Forge Press.
Morris, D. (1998). *Illness and culture in the postmodern age.* University of California Press.
Mouzelis, N. (1995). *Sociological theory: What went wrong? Diagnosis and remedies.* Routledge.
Mullender, A., & Hague, G. (2005). Giving voice to women survivors of domestic violence through recognition as a service user group. *British Journal of Social Work, 35*(8), 1321–1341.
Neugarten, D. (Ed.). (1996). *The meanings of age.* University of Chicago Press.
Newman, J., Glendinning, C., & Hughes, M. (2008). Beyond modernisation? Social care and the transformation of welfare governance. *Journal of Social Policy, 37*(4), 531–557.
O'Keeffe, M., Hills, A., Doyle, M., McCreadie, C., Scholes, S., Constantine, R., Tinker, A., Manthorpe, J., Biggs, S., & Erens, B. (2007). *UK study of abuse and neglect of older people: Prevalence survey report.* Prepared for comic relief and the Department of Health. National Centre for Social Research/King's College London.
Organization for Economic Cooperation and Development (OECD) Directorate for Employment, Labour and Social Affairs. (2007). *Disability trends among elderly people: Re-assessing the evidence in 12 OECD countries (interim report).* OECD.
Osborne, T. (1997). Of health and statecraft. In A. Petersen & R. Bunton (Eds.), *Foucault: Health and medicine* (pp. 173–188). Routledge.
Payne, M. (2004). Social work practice identities: An agency study of a Hospice. *Social Work in Action, 16*(1), 5–15.
Penna, S., & O'Brien, M. (1998). *Theorising welfare enlightenment and modern society.* Sage.
Petersen, A. (1997). Risk, governance and the new public health. In A. Petersen & R. Bunton (Eds.), *Foucault: Health and medicine* (pp. 189–206). Routledge.
Phillips, D. A. (2004). Reproducing normative and marginalised masculinities: Adolescent male popularity and the outcast. *Nursing Inquiry, 12*(3), 219–230.
Phillipson, C. (1982). *Capitalism and the construction of old age.* Macmillan.
Phillipson, C. (1986). Social policy and the older worker. In C. Phillipson & A. Walker (Eds.), *Ageing and social policy: A critical assessment* (pp. 45–60). Gower Publishing.
Phillipson, C. (1998). *Reconstructing old age.* Sage.
Phillipson, C. (2013). *Ageing.* Cambridge University Press.
Phillipson, C., & Powell, J. L. (2004). Risk, social welfare and old age. In E. Tulle (Ed.), *Old age and human agency.* Nova Science.
Phillipson, C., Bernard, M., Phillips, J., & Ogg, J. (2000). *The family and community life of career people.* Routledge.
Pillemer, K., & Wolf, R. (1986). *Elder abuse: Conflict in the family.* Auburn House.
Potter, J. (2001). Wittgenstein and Austin. In M. Wetherell, S. Taylor, & S. J. Yates (Eds.), *Discourse theory and practice.* Sage.
Powell, J. (1998a). Review of Katz, S (1996) 'Disciplining old age: The formation of gerontological knowledge'. *Acta Sociologica, 41*(4), 398–399.

Powell, J. (1998b). *The us and the them: Connecting Foucauldian and political economy insights into ageing bodies*. Unpublished paper presented British Sociological Association's annual conference, University of Edinburgh.

Powell, J. (1999). The importance of a 'critical' sociology of old age. *Social Science Paper Publisher, 3*, 1.

Powell, J. L. (2001). Theorizing gerontology: The case of old age, professional power and social policy in the United Kingdom. *Journal of Aging and Identity, 6*(3), 117–135.

Powell, J. L. (2001a). Theorizing gerontology: The case of old age, professional power and social policy in the United Kingdom. *Journal of Aging & Identity, 6*(3), 117–135.

Powell, J. L. (2001b). The NHS and Community Care Act (1990) in the UK: A critical review. *Sincronia: Journal of Social Sciences and Humanities, 5*(3), 1–10.

Powell, J. L. (2002). 'Understanding Habermas: Modern Solutions, postmodern problems. *Sincronia: Journal of Social Sciences and Humanities, 6*(2), 1–10.

Powell, J. (2005). *Social theory and aging*. Rowman and Littlefield.

Powell, J. (2006a). *Rethinking social theory and later life*. Nova Science.

Powell, J. (2006b). *Social theory and aging*. Rowman and Littlefield.

Powell, J. (2011a). *Aging, theory and globalization*. Nova Science.

Powell, J. (2011b). *Aging and social policy; a Foucauldian excursion*. Nova Science.

Powell, J. (2017). *Rethinking aging and death*. Nova Science.

Powell, J. L., & Biggs, S. (2000). Managing Old Age: The disciplinary web of power, Surveillance and Normalisation. *Journal of Aging and Identity, 5*(1), 3–13.

Powell, J. L., & Biggs, S. (2002). Foucauldian gerontology: A methodology for researching aging. *Electronic Journal of Sociology, 7*(2), 1–10.

Powell, J. L., & Biggs, S. (2003). Ageing, technologies of self, and bio-medicine: A Foucauldian excursion. *International Journal of Sociology and Social Policy, 23*(13), 96–115.

Powell, J., & Chamberlain, M. (2012). *Social welfare, aging and social theory*. Lexington.

Powell, J. L., & Cook, I. (2000). A tiger behind and coming up fast: Governmentality and the politics of population control. *Journal of Aging and Identity, 5*(2), 79–91.

Powell, J. L., & Gilbert, T. (2011). Personalisation and sustainable care. *Journal of Care Services Management, 5*(2), 79–86.

Powell, J. L., & Longino, C. F. (2001). Towards the postmodernization of aging: The body and social theory. *Journal of Aging and Identity, 6*(4), 20–34.

Powell, J. L., & Longino, C. F. (2002). Postmodernism V modernism: Rethinking theoretical tensions in social gerontology. *Journal of Aging and Identity, 7*(4), 97–107.

Powell, J., & Owen, T. (2007). *Reconstructing postmodernism: Critical debates*. Nova Science.

Powell, J., & Wahdin, A. (2004). Ageing and institutional abuse: A Foucauldian understanding. *International Journal of Sociology and Social Policy, 24*(12), 45–66.

Pruner, H. (1978). *Freud: His life and his mind*. Routledge.

Qualls, S. (1999). Realising power in intergenerational hierarchies. In M. Duffy (Ed.), *Handbook of counselling and psychotherapy with career adults*. Wiley.

Rabinow, P. (Ed.). (1984). *The Foucault reader*. Peregrine.

Reddy, P. (1996). Public service transformation and affirmative action perspectives in South Africa. *Public Personnel Management., 25*(1), 44–62.

Roberto, K. (1990). Grandparent and grand-child relationships. In T. H. Brubaker (Ed.), *Family relationships in later life*. Shealth and social care.

Rose, N. (1984). Genetic counseling: The development of a new interdisciplinary occupational field. *Social Science and Medicine, 18*(7), 541–549.

Rose, N. (1993, August). Government, authority and expertise in advanced liberalism. *Economy and Society, 22*(3), 283–299.

Rose, N. (1996). The death of the social? Re-figuring the territory of government. *Economy and Society, 25*(3), 327–356.

Rose, N. (1999). *Powers of freedom: Reframing political thought*. Cambridge University Press.

Rose, N., & Miller, P. (1992). Political power beyond the state: Problematics of government. *The British Journal of Sociology, 43*(2), 173–205.

Rouse, J. (1994). Power/knowledge. In G. Gutting (Ed.), *The Cambridge companion to Foucault*. Cambridge University Press.

Ryff, C., & Seltzer, M. (1996). *The parental experience in midlife*. Chicago University Press.

Samuel, M. (2008, February 20). No secrets review: The key issues. *Community Care*.

Samuel, M. (2009, March 18). Practitioners back personalisation but call for more support. *Community Care*.

Satyamurti, C. (1974). *Women's occupation and social change: The case of social work*. Paper presented to 1974 British Sociological Association's annual conference.

Scharf, T., & Wenger, G. (1995). *International perspectives on community care for career people*. Avebury.

Schreck, H. (2000). *Community and caring*. UPA.

Schryer, C. F., & Spoel, P. (2005). Genre theory, health-care discourse, and professional identity formation. *Journal of Business and Technical communication, 19*(3), 249–278.

Searle, J. H. (1969). *Speech acts: An essay in the philosophy of language*. Cambridge University Press.

Sheppard, M. (1995a). Social work, social science and practice wisdom. *British Journal of Social Work, 25*, 265–293.

Sheppard, M. (1995b). *Care management and the new social work*. Whiting and Birch.

Shumway, D. (1989). *Michel Foucault*. University of Virginia Press.

Silverstein, M., & Bengtson, V. L. (1997). Intergenerational solidarity and the structure of adult child-parent relationships in American families. *American Journal of Sociology, 103*(2), 429–460.

Smale, G., Tuson, G., & Statham, D. (2000). *Social work and social problems*. Macmillan.

Smart, B. (1983). *Foucault, Marxism and critique*. Routledge and Kegan Paul.

Smart, B. (1985). *Michel Foucault*. Routledge.

Smith, H., & Brown, H. (1992). Defending community care: Can normalisation do the job? *British Journal of Social Work, 22*(6), 685–693.

Sturdy, A., & Fleming, P. (2003). Talk as technique – A critique of the words and deeds distinction in the diffusion of customer service cultures in call centres. *Journal of Management Studies, 40*(4), 753–774.

Taylor-Gooby, P. (2000). Risk and welfare. In P. Taylor-Gooby (Ed.), *Risk, trust and welfare* (pp. 1–20). Macmillan.

The National Assembly for Wales and the Home Office. (2000). *In safe hands: Implementing adult protection procedures in Wales*. The National Assembly for Wales.

Thompson, P. (1999). The role of grandparents when parents part or die: Some reflections on the mythical decline of the extended family. *Health and Social Careing and Society, 19*(4), 471–503.

Thompson, N. (2000). *Understanding social work: Preparing for practice*. Macmillan.

Thompson, P., & Ackroyd, S. (1995). All quiet on the workplace front? A critique of recent trends in British industrial sociology. *Sociology, 29*(4), 615–633.

Thomson, E., & Minkler, M. (2001). American grandparents providing extensive childcare to their grand-children: Prevalence and profile. *The Gerontologist, 41*(2), 201–209.

Timiras, P. (1997). *Advances in cell aging and gerontology*. Science Direct Press.

Townsend, P. (1981). The structured dependency of the elderly: A creation of social policy in the twentieth century. *Ageing and Society, 1*(1), 5–28.

Townsend, P. (1986). Ageism and social policy. In C. Phillipson & A. Walker (Eds.), *Ageing and social policy*. Aldershot.

Turner, B. (1989). Ageing, status politics and sociological theory. *British Journal of Sociology, 40*, 588–606.

Turner, B. S. (1997). From governmentality to risk: Some reflections on Foucault's contribution to medical sociology. In A. Petersen & R. Bunton (Eds.), *Foucault: Health and medicine* (pp. ix–xxi). Routledge.

Turner, B., & Dumas, A. (2006). Age and aging: the social world of Foucault and Bourdieu. In J. Powell & A. Wahidin (Eds.), *Foucault and aging*. Nova Science.

UK Government (UKG). (1981). *Growing old*. HMSO.
UK Government (UKG). (1989). *Community care: An agenda for action*. HMSO.
UK Government (UKG). (1990). *NHS and community care act*. HMSO.
UK Government (UKG). (2000a). *Winning the generation game*. www.cabinet-office.gov.uk
UK Government (UKG). (2000b). *Supporting families*. HMSO.
United Nations Department of Economic and Social Affairs. (2002). *Population division. World population aging 1950–2050*. United Nations.
Vincent, J. (1996). Whose afraid of an ageing population? *Critical Social Policy., 47*, 3–26.
Virilio, P. (1983). *Pure war*. MIT Press.
Wahidin, A., & Powell, J. (2001). The loss of Aging identity: Social theory and the power of special hospitals. *Journal of Aging & Identity, 6*(1), 31–49.
Waldrop, D., Weber, J., Herald, S., Pruett, J., Cooper, K., & Jouzapavicius, K. (1999). Wisdom and life experience: How grandfathers mentor their grandchildren. *Journal of Aging and Identity, 4*(1), 33–46.
Walker, A. (1981). Towards a political economy of old age. *Ageing & Society, 1*(1), 73–94.
Walker, A. (2002). A strategy for active ageing. *International Health Security Review, 55*(1), 121–139.
Walker, A., & Aspalter, C. (Eds.). (2008). *Securing the future for older people in Europe*. Casa Verde.
Walker, A., & Naegele, G. (1999). *The politics of old age in Europe*. Open University Press.
Wetherell, M. (2001a). Themes in discourse research. In M. Wetherell, S. Taylor, & S. J. Yates (Eds.), *Discourse theory and practice*. Sage.
Wetherell, M. (2001b). Themes in discourse research: The case of Diana. In M. Wetherell, S. Taylor, & S. J. Yates (Eds.), *Discourse theory and practice* (pp. 14–28). Sage.
Whitehead, S. (1998). Disrupted selves: Resistance and identity work in the managerial arena. *Gender and Education, 10*(2), 199–215.
Wilkinson, I. (2001). *Anxiety in a risk society*. Routledge.
Williams, F. (1992). Somewhere over the rainbow: Universality and diversity in social policy. In N. Manning & R. Page (Eds.), *Social Policy Review* (Vol. 4). SPA.
Wilson, G. (1991). Models of Ageing and their relation to policy formulation. *Policy & Politics, 19*(1), 37–47.

Index

J. Powell, *New Perspectives on Health and Social Care*, International Perspectives on Social Policy, Administration, and Practice,
https://doi.org/10.1007/978-3-031-25432-1